W9-DBC-064

Undergraduate Texts in Mathematics

Undergraduate Texts in Mathematics

Abbott: Understanding Analysis.
Anglin: Mathematics: A Concise History and Philosophy.
Readings in Mathematics.
Anglin/Lambek: The Heritage of Thales.
Readings in Mathematics.
Apostol: Introduction to Analytic Number Theory. Second edition.
Armstrong: Basic Topology.
Armstrong: Groups and Symmetry.
Axler: Linear Algebra Done Right. Second edition.
Beardon: Limits: A New Approach to Real Analysis.
Bak/Newman: Complex Analysis. Second edition.
Banchoff/Wermer: Linear Algebra Through Geometry. Second edition.
Berberian: A First Course in Real Analysis.
Bix: Conics and Cubics: A Concrete Introduction to Algebraic Curves.
Brémaud: An Introduction to Probabilistic Modeling.
Bressoud: Factorization and Primality Testing.
Bressoud: Second Year Calculus.
Readings in Mathematics.
Brickman: Mathematical Introduction to Linear Programming and Game Theory.
Browder: Mathematical Analysis: An Introduction.
Buchmann: Introduction to Cryptography.
Buskes/van Rooij: Topological Spaces: From Distance to Neighborhood.
Callahan: The Geometry of Spacetime: An Introduction to Special and General Relavitity.
Carter/van Brunt: The Lebesgue–Stieltjes Integral: A Practical Introduction.
Cederberg: A Course in Modern Geometries. Second edition.
Chambert-Loir: A Field Guide to Algebra
Childs: A Concrete Introduction to Higher Algebra. Second edition.
Chung/AitSahlia: Elementary Probability Theory: With Stochastic Processes and an Introduction to Mathematical Finance. Fourth edition.
Cox/Little/O'Shea: Ideals, Varieties, and Algorithms. Second edition.
Croom: Basic Concepts of Algebraic Topology.
Cull/Flahive/Robson: Difference Equations: From Rabbits to Chaos
Curtis: Linear Algebra: An Introductory Approach. Fourth edition.
Daepp/Gorkin: Reading, Writing, and Proving: A Closer Look at Mathematics.
Devlin: The Joy of Sets: Fundamentals of Contemporary Set Theory. Second edition.
Dixmier: General Topology.
Driver: Why Math?
Ebbinghaus/Flum/Thomas: Mathematical Logic. Second edition.
Edgar: Measure, Topology, and Fractal Geometry.
Elaydi: An Introduction to Difference Equations. Third edition.
Erdõs/Surányi: Topics in the Theory of Numbers.
Estep: Practical Analysis in One Variable.
Exner: An Accompaniment to Higher Mathematics.
Exner: Inside Calculus.
Fine/Rosenberger: The Fundamental Theory of Algebra.
Fischer: Intermediate Real Analysis.
Flanigan/Kazdan: Calculus Two: Linear and Nonlinear Functions. Second edition.
Fleming: Functions of Several Variables. Second edition.
Foulds: Combinatorial Optimization for Undergraduates.
Foulds: Optimization Techniques: An Introduction.

(continued on page 228)

John Stillwell

The Four Pillars of Geometry

With 138 Illustrations

Springer

John Stillwell
Department of Mathematics
University of San Francisco
San Francisco, CA 94117-1080
USA
stillwell@usfca.edu

Mathematics Subject Classification (2000): 51-xx, 15-xx

Library of Congress Control Number: 2005929630

ISBN-10: 0-387-25530-3
ISBN-13: 978-0387-25530-9

Printed on acid-free paper.

Printed in the United States of America. (EB)

9 8 7 6 5 4 3 2 1

springeronline.com

To Elaine

Preface

Many people think there is only one "right" way to teach geometry. For two millennia, the "right" way was Euclid's way, and it is still good in many respects. But in the 1950s the cry "Down with triangles!" was heard in France and new geometry books appeared, packed with linear algebra but with no diagrams. Was this the new "right" way, or was the "right" way something else again, perhaps transformation groups?

In this book, I wish to show that geometry can be developed in four fundamentally different ways, and that *all* should be used if the subject is to be shown in all its splendor. Euclid-style construction and axiomatics seem the best way to start, but linear algebra smooths the later stages by replacing some tortuous arguments by simple calculations. And how can one avoid projective geometry? It not only explains why objects look the way they do; it also explains why geometry is entangled with algebra. Finally, one needs to know that there is not one geometry, but many, and transformation groups are the best way to distinguish between them.

Two chapters are devoted to each approach: The first is concrete and introductory, whereas the second is more abstract. Thus, the first chapter on Euclid is about straightedge and compass constructions; the second is about axioms and theorems. The first chapter on linear algebra is about coordinates; the second is about vector spaces and the inner product. The first chapter on projective geometry is about perspective drawing; the second is about axioms for projective planes. The first chapter on transformation groups gives examples of transformations; the second constructs the hyperbolic plane from the transformations of the real projective line.

I believe that students are shortchanged if they miss any of these four approaches to the subject. Geometry, of all subjects, should be about *taking different viewpoints*, and geometry is unique among the mathematical disciplines in its ability to look different from different angles. Some prefer to approach it visually, others algebraically, but the miracle is that they are all looking at the same thing. (It is as if one discovered that number theory need not use addition and multiplication, but could be based on, say, the exponential function.)

The many faces of geometry are not only a source of amazement and delight. They are also a great help to the learner and teacher. We all know that some students prefer to visualize, whereas others prefer to reason or to calculate. Geometry has something for everybody, and all students will find themselves building on their strengths at some times, and working to overcome weaknesses at other times. We also know that Euclid has some beautiful proofs, whereas other theorems are more beautifully proved by algebra. In the multifaceted approach, every theorem can be given an elegant proof, and theorems with radically different proofs can be viewed from different sides.

This book is based on the course Foundations of Geometry that I taught at the University of San Francisco in the spring of 2004. It should be possible to cover it all in a one-semester course, but if time is short, some sections or chapters can be omitted according to the taste of the instructor. For example, one could omit Chapter 6 or Chapter 8. (But with regret, I am sure!)

Acknowledgements

My thanks go to the students in the course, for feedback on my raw lecture notes, and especially to Gina Campagna and Aaron Keel, who contributed several improvements.

Thanks also go to my wife Elaine, who proofread the first version of the book, and to Robin Hartshorne, John Howe, Marc Ryser, Abe Shenitzer, and Michael Stillwell, who carefully read the revised version and saved me from many mathematical and stylistic errors.

Finally, I am grateful to the M. C. Escher Company – Baarn – Holland for permission to reproduce the Escher work *Circle Limit I* shown in Figure 8.19, and the explicit mathematical transformation of it shown in Figure 8.10.

JOHN STILLWELL
San Francisco, November 2004
South Melbourne, April 2005

Contents

1

Straightedge and compass

PREVIEW

For over 2000 years, mathematics was almost synonymous with the geometry of Euclid's *Elements*, a book written around 300 BCE and used in school mathematics instruction until the 20th century. *Euclidean geometry*, as it is now called, was thought to be the foundation of all exact science.

Euclidean geometry plays a different role today, because it is no longer expected to support everything else. "Non-Euclidean geometries" were discovered in the early 19th century, and they were found to be more useful than Euclid's in certain situations. Nevertheless, non-Euclidean geometries arose as deviations from the Euclidean, so one first needs to know *what* they deviate from.

A naive way to describe Euclidean geometry is to say it concerns the geometric figures that can be drawn (or *constructed* as we say) by straightedge and compass. Euclid assumes that it is possible to draw a straight line between any two given points, and to draw a circle with given center and radius. All of the propositions he proves are about figures built from straight lines and circles.

Thus, to understand Euclidean geometry, one needs some idea of the scope of straightedge and compass constructions. This chapter reviews some basic constructions, to give a quick impression of the extent of Euclidean geometry, and to suggest why *right angles* and *parallel lines* play a special role in it.

Constructions also help to expose the role of length, area, and angle in geometry. The deeper meaning of these concepts, and the related role of *numbers* in geometry, is a thread we will pursue throughout the book.

1.1 Euclid's construction axioms

Euclid assumes that certain constructions can be done and he states these assumptions in a list called his *axioms* (traditionally called *postulates*). He assumes that it is possible to:

1. Draw a straight line segment between any two points.

2. Extend a straight line segment indefinitely.

3. Draw a circle with given center and radius.

Axioms 1 and 2 say we have a *straightedge*, an instrument for drawing arbitrarily long line segments. Euclid and his contemporaries tried to avoid infinity, so they worked with line segments rather than with whole lines. This is no real restriction, but it involves the annoyance of having to extend line segments (or "produce" them, as they say in old geometry books). Today we replace Axioms 1 and 2 by the single axiom that a *line* can be drawn through any two points.

The straightedge (unlike a ruler) has no scale marked on it and hence can be used *only* for drawing lines—not for measurement. Euclid separates the function of measurement from the function of drawing straight lines by giving measurement functionality only to the *compass*—the instrument assumed in Axiom 3. The compass is used to draw the circle through a given point B, with a given point A as center (Figure 1.1).

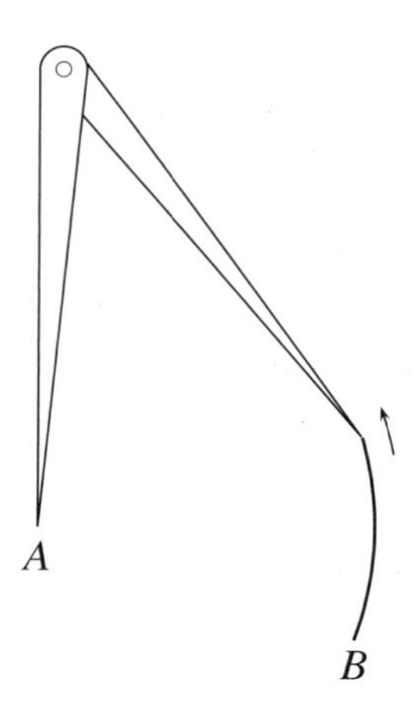

Figure 1.1: Drawing a circle

To do this job, the compass must rotate rigidly about A after being initially set on the two points A and B. Thus, it "stores" the length of the radius AB and allows this length to be transferred elsewhere. Figure 1.2 is a classic view of the compass as an instrument of measurement. It is William Blake's painting of Isaac Newton as the measurer of the universe.

Figure 1.2: Blake's painting of Newton the measurer

The compass also enables us to *add* and *subtract* the length $|AB|$ of AB from the length $|CD|$ of another line segment CD by picking up the compass with radius set to $|AB|$ and describing a circle with center D (Figure 1.3, also *Elements*, Propositions 2 and 3 of Book I). By adding a fixed length repeatedly, one can construct a "scale" on a given line, effectively creating a ruler. This process illustrates how the power of measuring lengths resides in the compass. Exactly which lengths can be measured in this way is a deep question, which belongs to algebra and analysis. The full story is beyond the scope of this book, but we say more about it below.

Separating the concepts of "straightness" and "length," as the straightedge and the compass do, turns out to be important for understanding the foundations of geometry. The same separation of concepts reappears in different approaches to geometry developed in Chapters 3 and 5.

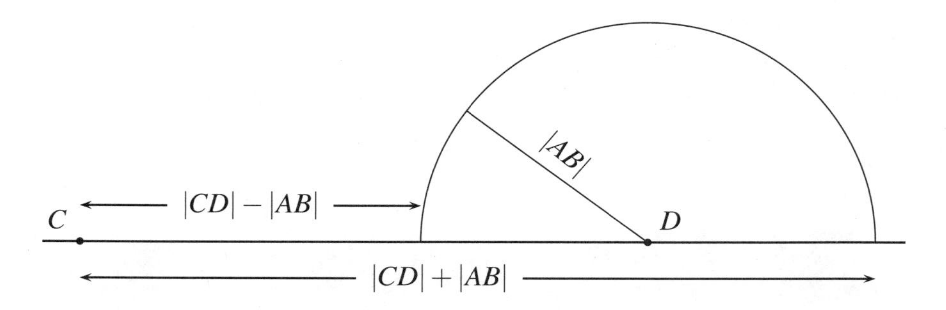

Figure 1.3: Adding and subtracting lengths

1.2 Euclid's construction of the equilateral triangle

Constructing an equilateral triangle on a given side AB is the first proposition of the *Elements*, and it takes three steps:

1. Draw the circle with center A and radius AB.
2. Draw the circle with center B and radius AB.
3. Draw the line segments from A and B to the intersection C of the two circles just constructed.

The result is the triangle ABC with sides AB, BC, and CA in Figure 1.4.

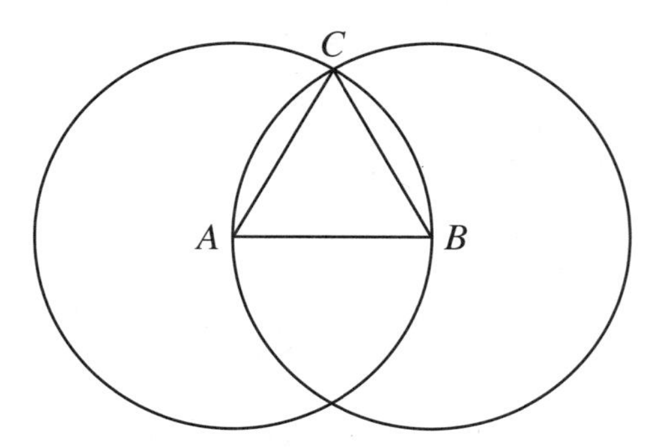

Figure 1.4: Constructing an equilateral triangle

Sides AB and CA have equal length because they are both radii of the first circle. Sides AB and BC have equal length because they are both radii of the second circle. Hence, all three sides of triangle ABC are equal. □

This example nicely shows the interplay among

- *construction axioms*, which guarantee the existence of the construction lines and circles (initially the two circles on radius AB and later the line segments BC and CA),

- *geometric axioms*, which guarantee the existence of points required for later steps in the construction (the intersection C of the two circles),

- and *logic*, which guarantees that certain conclusions follow. In this case, we are using a principle of logic that says that things equal to the same thing (both $|BC|$ and $|CA|$ equal $|AB|$) are equal to each other (so $|BC| = |CA|$).

We have not yet discussed Euclid's geometric axioms or logic. We use the same logic for all branches of mathematics, so it can be assumed "known," but geometric axioms are less clear. Euclid drew attention to one and used others unconsciously (or, at any rate, without stating them). History has shown that Euclid correctly identified the most significant geometric axiom, namely the *parallel axiom*. We will see some reasons for its significance in the next section. The ultimate reason is that *there are important geometries in which the parallel axiom is false*.

The other axioms are not significant in this sense, but they should also be identified for completeness, and we will do so in Chapter 2. In particular, it should be mentioned that Euclid states no axiom about the intersection of circles, so he has not justified the existence of the point C used in his very first proposition!

A question arising from Euclid's construction

The equilateral triangle is an example of a *regular polygon*: a geometric figure bounded by equal line segments that meet at equal angles. Another example is the regular hexagon in Exercise 1.2.1. If the polygon has n sides, we call it an *n-gon*, so the regular 3-gon and the regular 6-gon are constructible. *For which n is the regular n-gon constructible?*

We will not completely answer this question, although we will show that the regular 4-gon and 5-gon are constructible. The question for general n turns out to belong to algebra and number theory, and a complete answer depends on a problem about prime numbers that has not yet been solved: For which m is $2^{2^m} + 1$ a prime number?

Exercises

By extending Euclid's construction of the equilateral triangle, construct:

1.2.1 A regular hexagon.

1.2.2 A tiling of the plane by equilateral triangles (solid lines in Figure 1.5).

1.2.3 A tiling of the plane by regular hexagons (dashed lines in Figure 1.5).

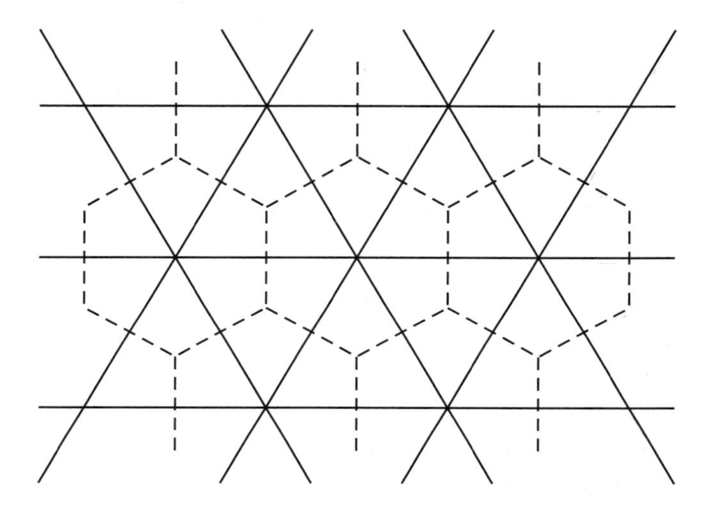

Figure 1.5: Triangle and hexagon tilings of the plane

1.3 Some basic constructions

The equilateral triangle construction comes first in the *Elements* because several other constructions follow from it. Among them are constructions for bisecting a line segment and bisecting an angle. ("Bisect" is from the Latin for "cut in two.")

Bisecting a line segment

To bisect a given line segment AB, draw the two circles with radius AB as above, but now consider both of their intersection points, C and D. The line CD connecting these points bisects the line segment AB (Figure 1.6).

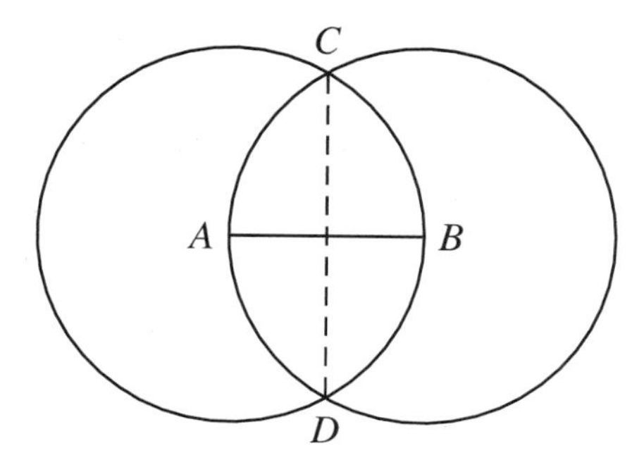

Figure 1.6: Bisecting a line segment AB

Notice also that BC is *perpendicular* to AB, so this construction can be adapted to construct perpendiculars.

- To construct the perpendicular to a line $\mathscr{L}$ at a point E on the line, first draw a circle with center E, cutting $\mathscr{L}$ at A and B. Then the line CD constructed in Figure 1.6 is the perpendicular through E.

- To construct the perpendicular to a line $\mathscr{L}$ through a point E not on $\mathscr{L}$, do the same; only make sure that the circle with center E is large enough to cut the line $\mathscr{L}$ at two different points.

Bisecting an angle

To bisect an angle POQ (Figure 1.7), first draw a circle with center O cutting OP at A and OQ at B. Then the perpendicular CD that bisects the line segment AB also bisects the angle POQ.

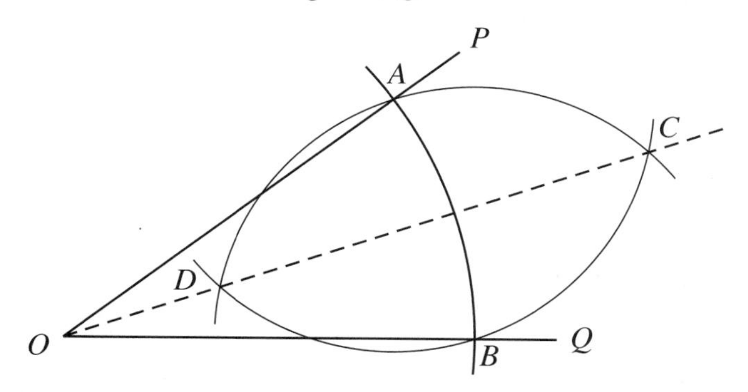

Figure 1.7: Bisecting an angle POQ

It seems from these two constructions that bisecting a line segment and bisecting an angle are virtually the same problem. Euclid bisects the angle before the line segment, but he uses two similar constructions (*Elements*, Propositions 9 and 10 of Book I). However, a distinction between line segments and angles emerges when we attempt division into three or more parts. There is a simple tool for dividing a line segment in any number of equal parts—*parallel lines*—but no corresponding tool for dividing angles.

Constructing the parallel to a line through a given point

We use the two constructions of perpendiculars noted above—for a point off the line and a point on the line. Given a line $\mathscr{L}$ and a point P outside $\mathscr{L}$, first construct the perpendicular line $\mathscr{M}$ to $\mathscr{L}$ through P. Then construct the perpendicular to $\mathscr{M}$ through P, which is the parallel to $\mathscr{L}$ through P.

Dividing a line segment into n equal parts

Given a line segment AB, draw any other line $\mathscr{L}$ through A and mark n successive, equally spaced points $A_1, A_2, A_3, \ldots, A_n$ along $\mathscr{L}$ using the compass set to any fixed radius. Figure 1.8 shows the case $n = 5$. Then connect A_n to B, and draw the parallels to BA_n through $A_1, A_2, \ldots, A_{n-1}$. These parallels divide AB into n equal parts.

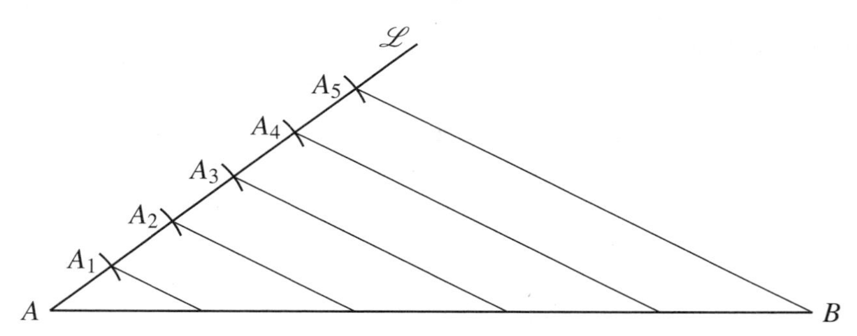

Figure 1.8: Dividing a line segment into equal parts

This construction depends on a property of parallel lines sometimes attributed to Thales (Greek mathematician from around 600 BCE): *parallels cut any lines they cross in proportional segments*. The most commonly used instance of this theorem is shown in Figure 1.9, where a parallel to one side of a triangle cuts the other two sides proportionally.

The line $\mathscr{L}$ parallel to the side BC cuts side AB into the segments AP and PB, side AC into AQ and QC, and $|AP|/|PB| = |AQ|/|QC|$.

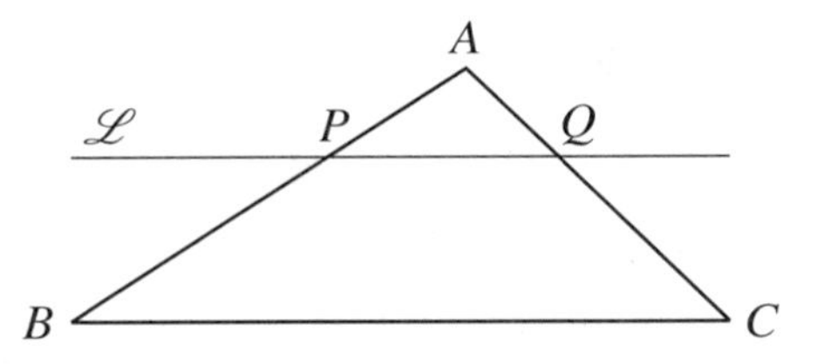

Figure 1.9: The Thales theorem in a triangle

This theorem of Thales is the key to using algebra in geometry. In the next section we see how it may be used to multiply and divide line segments, and in Chapter 2 we investigate how it may be derived from fundamental geometric principles.

Exercises

1.3.1 Check for yourself the constructions of perpendiculars and parallels described in words above.

1.3.2 Can you find a more direct construction of parallels?

Perpendiculars give another important polygon—the square.

1.3.3 Give a construction of the square on a given line segment.

1.3.4 Give a construction of the square tiling of the plane.

One might try to use division of a line segment into n equal parts to divide an angle into n equal parts as shown in Figure 1.10. We mark A on OP and B at equal distance on OQ as before, and then try to divide angle POQ by dividing line segment AB. However, this method is faulty even for division into three parts.

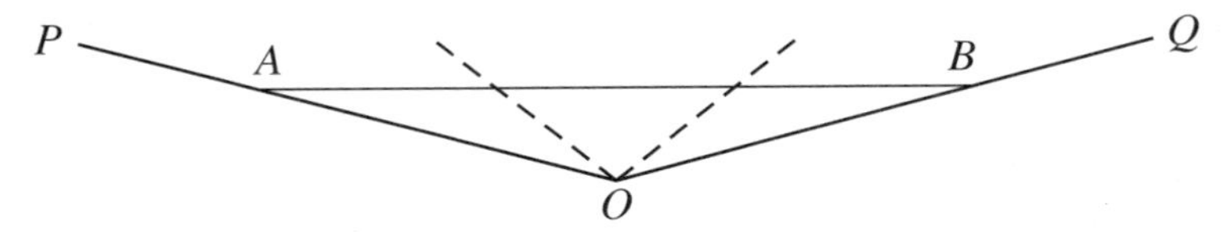

Figure 1.10: Faulty trisection of an angle

1.3.5 Explain why division of AB into three equal parts (trisection) does *not* always divide angle POQ into three equal parts. (Hint: Consider the case in which POQ is nearly a straight line.)

The version of the Thales theorem given above (referring to Figure 1.9) has an equivalent form that is often useful.

1.3.6 If A, B, C, P, Q are as in Figure 1.9, so that $|AP|/|PB| = |AQ|/|QC|$, show that this equation is equivalent to $|AP|/|AB| = |AQ|/|AC|$.

1.4 Multiplication and division

Not only can one add and subtract line segments (Section 1.1); one can also multiply and divide them. The *product* ab and *quotient* a/b of line segments a and b are obtained by the straightedge and compass constructions below. The key ingredients are parallels, and the key geometric property involved is the Thales theorem on the proportionality of line segments cut off by parallel lines.

To get started, it is necessary to choose a line segment as the *unit of length*, 1, which has the property that $1a = a$ for any length a.

Product of line segments

To multiply line segment b by line segment a, we first construct any triangle UOA with $|OU| = 1$ and $|OA| = a$. We then extend OU by length b to B_1 and construct the parallel to UA through B_1. Suppose this parallel meets the extension of OA at C (Figure 1.11).

By the Thales theorem, $|AC| = ab$.

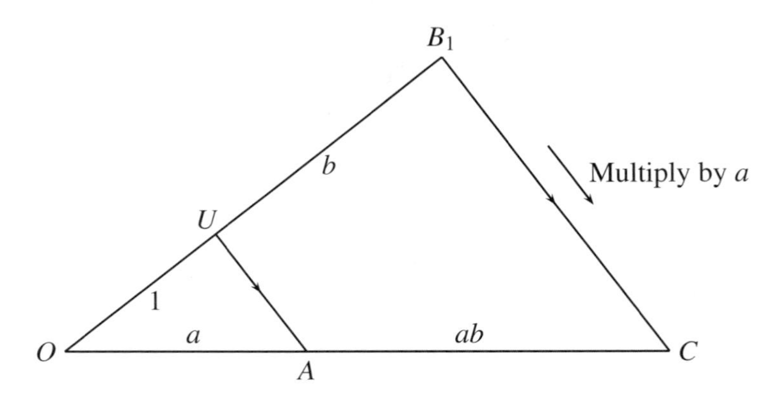

Figure 1.11: The product of line segments

Quotient of line segments

To divide line segment b by line segment a, we begin with the same triangle UOA with $|OU| = 1$ and $|OA| = a$. Then we extend OA by distance b to B_2 and construct the parallel to UA through B_2. Suppose that this parallel meets the extension of OU at D (Figure 1.12).

By the Thales theorem, $|UD| = b/a$.

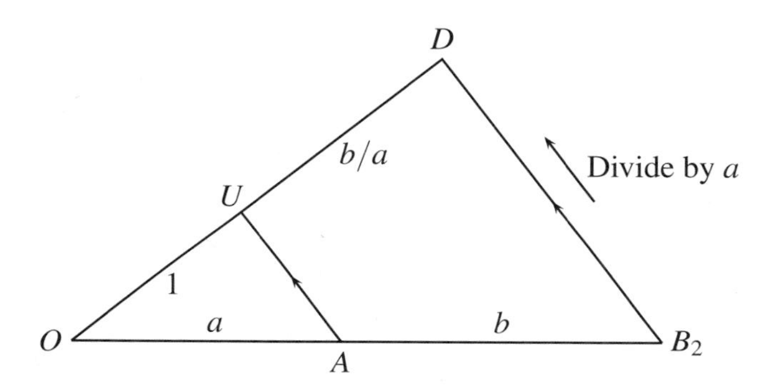

Figure 1.12: The quotient of line segments

The sum operation from Section 1.1 allows us to construct a segment n units in length, for any natural number n, simply by adding the segment 1 to itself n times. The quotient operation then allows us to construct a segment of length m/n, for any natural numbers m and $n \neq 0$. These are what we call the *rational* lengths. A great discovery of the Pythagoreans was that *some lengths are not rational*, and that some of these "irrational" lengths can be constructed by straightedge and compass. It is not known how the Pythagoreans made this discovery, but it has a connection with the Thales theorem, as we will see in the next section.

Exercises

Exercise 1.3.6 showed that if PQ is parallel to BC in Figure 1.9, then $|AP|/|AB| = |AQ|/|AC|$. That is, a parallel implies proportional (left and right) sides. The following exercise shows the converse: proportional sides imply a parallel, or (equivalently), a nonparallel implies nonproportional sides.

1.4.1 Using Figure 1.13, or otherwise, show that if PR is *not* parallel to BC, then $|AP|/|AB| \neq |AR|/|AC|$.

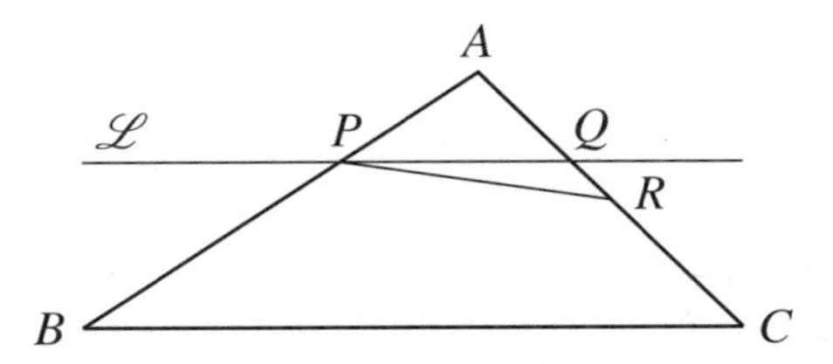

Figure 1.13: Converse of the Thales theorem

1.4.2 Conclude from Exercise 1.4.1 that if P is any point on AB and Q is any point on AC, then PQ is parallel to BC *if and only if* $|AP|/|AB| = |AQ|/|AC|$.

The "only if" direction of Exercise 1.4.2 leads to two famous theorems—the *Pappus* and *Desargues theorems*—that play an important role in the foundations of geometry. We will meet them in more general form later. In their simplest form, they are the following theorems about parallels.

1.4.3 (Pappus of Alexandria, around 300 CE) Suppose that A, B, C, D, E, F lie alternately on lines $\mathscr{L}$ and $\mathscr{M}$ as shown in Figure 1.14.

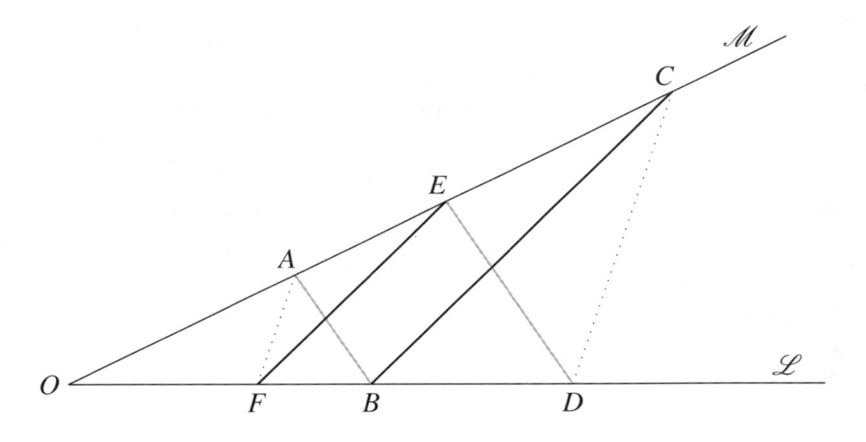

Figure 1.14: The parallel Pappus configuration

Use the Thales theorem to show that if AB is parallel to ED and FE is parallel to BC then

$$\frac{|OA|}{|OF|} = \frac{|OC|}{|OD|}.$$

Deduce from Exercise 1.4.2 that AF is parallel to CD.

1.4.4 (Girard Desargues, 1648) Suppose that points A,B,C,A',B',C' lie on concurrent lines $\mathscr{L},\mathscr{M},\mathscr{N}$ as shown in Figure 1.15. (The triangles ABC and $A'B'C'$ are said to be "in perspective from O.")

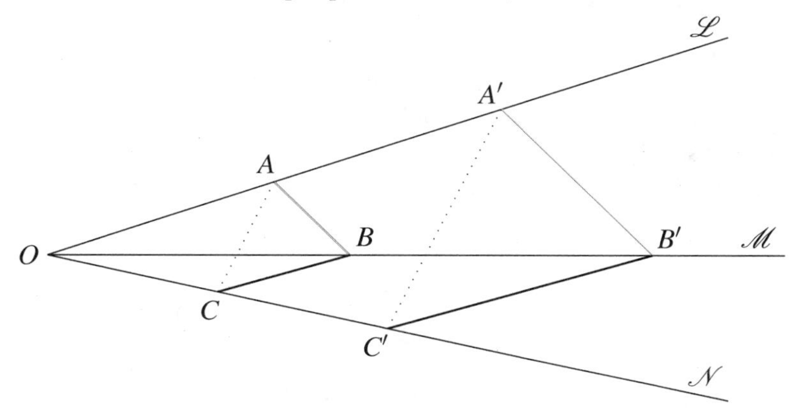

Figure 1.15: The parallel Desargues configuration

Use the Thales theorem to show that if AB is parallel to $A'B'$ and BC is parallel to $B'C'$, then

$$\frac{|OA|}{|OC|} = \frac{|OA'|}{|OC'|}.$$

Deduce from Exercise 1.4.2 that AC is parallel to $A'C'$.

1.5 Similar triangles

Triangles ABC and $A'B'C'$ are called *similar* if their corresponding angles are equal, that is, if

$$\begin{aligned}
\text{angle at } A &= \text{angle at } A' \quad (= \alpha \text{ say}),\\
\text{angle at } B &= \text{angle at } B' \quad (= \beta \text{ say}),\\
\text{angle at } C &= \text{angle at } C' \quad (= \gamma \text{ say}).
\end{aligned}$$

It turns out that equal angles imply that *all sides are proportional*, so we may say that one triangle is a magnification of the other, or that they have the same "shape." This important result extends the Thales theorem, and actually follows from it.

Why similar triangles have proportional sides

Imagine moving triangle ABC so that vertex A coincides with A' and sides AB and AC lie on sides $A'B'$ and $A'C'$, respectively. Then we obtain the situation shown in Figure 1.16. In this figure, b and c denote the side lengths of triangle ABC opposite vertices B and C, respectively, and b' and c' denote the side lengths of triangle $A'B'C'(=AB'C')$ opposite vertices B' and C', respectively.

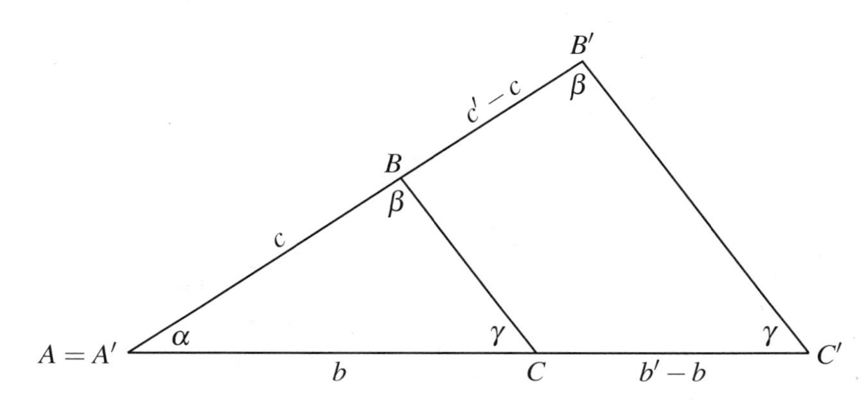

Figure 1.16: Similar triangles

Because BC and $B'C'$ both meet AB' at angle β, they are parallel, and so it follows from the Thales theorem (Section 1.3) that

$$\frac{b}{c} = \frac{b' - b}{c' - c}.$$

Multiplying both sides by $c(c'-c)$ gives $b(c'-c) = c(b'-b)$, that is,

$$bc' - bc = cb' - cb,$$

and hence

$$bc' = cb'.$$

Finally, dividing both sides by cc', we get

$$\frac{b}{c} = \frac{b'}{c'}.$$

That is, *corresponding sides of triangles ABC and $A'B'C'$ opposite to the angles β and γ are proportional.*

We got this result by making the angles α in the two triangles coincide. If we make the angles β coincide instead, we similarly find that the sides opposite to α and γ are proportional. Thus, in fact, *all corresponding sides of similar triangles are proportional.* □

This consequence of the Thales theorem has many implications. In everyday life, it underlies the existence of scale maps, house plans, engineering drawings, and so on. In pure geometry, its implications are even more varied. Here is just one, which shows why square roots and irrational numbers turn up in geometry.

The diagonal of the unit square is $\sqrt{2}$

The diagonals of the unit square cut it into four quarters, each of which is a triangle similar to the half square cut off by a diagonal (Figure 1.17).

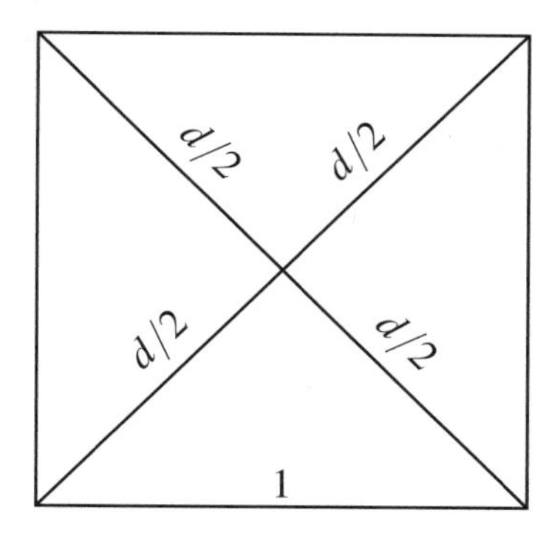

Figure 1.17: Quarters and halves of the square

Each of the triangles in question has one right angle and two half right angles, so it follows from the theorem above that corresponding sides of any two of these triangles are proportional. In particular, if we take the half square, with short side 1 and long side d, and compare it with the quarter square, with short side $d/2$ and long side 1, we get

$$\frac{\text{short}}{\text{long}} = \frac{1}{d} = \frac{d/2}{1}.$$

Multiplying both sides of the equation by $2d$ gives $2 = d^2$, so $d = \sqrt{2}$. □

The great, but disturbing, discovery of the Pythagoreans is that $\sqrt{2}$ *is irrational*. That is, there are no natural numbers m and n such $\sqrt{2} = m/n$.

If there are such m and n we can assume that they have no common divisor, and then the assumption $\sqrt{2} = m/n$ implies

	$2 = m^2/n^2$	squaring both sides
hence	$m^2 = 2n^2$	multiplying both sides by n^2
hence	m^2 is even	
hence	m is even	since the square of an odd number is odd
hence	$m = 2l$	for some natural number l
hence	$m^2 = 4l^2 = 2n^2$	
hence	$n^2 = 2l^2$	
hence	n^2 is even	
hence	n is even	since the square of an odd number is odd.

Thus, m and n have the common divisor 2, contrary to assumption. Our original assumption is therefore false, so there are no natural numbers m and n such that $\sqrt{2} = m/n$. □

Lengths, products, and area

Geometry obviously has to include the diagonal of the unit square, hence *geometry includes the study of irrational lengths.* This discovery troubled the ancient Greeks, because they did not believe that irrational lengths could be treated like numbers. In particular, the idea of interpreting the product of line segments as another line segment is *not* in Euclid. It first appears in Descartes' *Géométrie* of 1637, where algebra is used systematically in geometry for the first time.

The Greeks viewed the product of line segments a and b as the *rectangle* with perpendicular sides a and b. If lengths are not necessarily numbers, then the product of two lengths is best interpreted as an area, and the product of three lengths as a volume—but then the product of four lengths seems to have no meaning at all. This difficulty perhaps explains why algebra appeared comparatively late in the development of geometry. On the other hand, interpreting the product of lengths as an area gives some remarkable insights, as we will see in Chapter 2. So it is also possible that algebra had to wait until the Greek concept of product had exhausted its usefulness.

Exercises

In general, two geometric figures are called similar if one is a magnification of the other. Thus, two rectangles are similar if the ratio $\frac{\text{long side}}{\text{short side}}$ is the same for both.

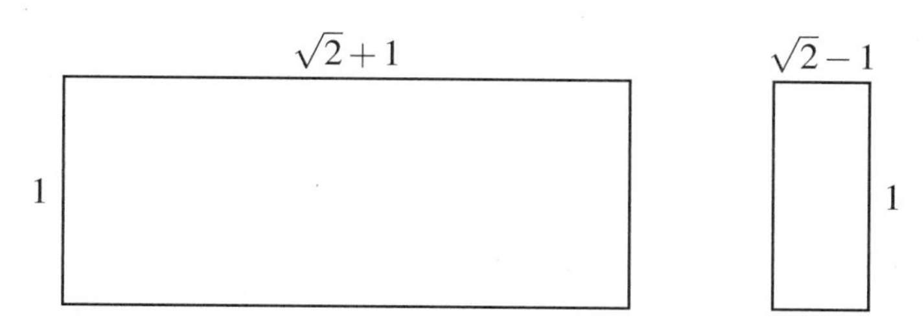

Figure 1.18: A pair of similar rectangles

1.5.1 Show that $\frac{\sqrt{2}+1}{1} = \frac{1}{\sqrt{2}-1}$ and hence that the two rectangles in Figure 1.18 are similar.

1.5.2 Deduce that if a rectangle with long side a and short side b has the same shape as the two above, then so has the rectangle with long side b and short side $a-2b$.

This simple observation gives another proof that $\sqrt{2}$ is irrational:

1.5.3 Suppose that $\sqrt{2}+1 = m/n$, where m and n are natural numbers with m as small as possible. Deduce from Exercise 1.5.2 that we also have $\sqrt{2}+1 = n/(m-2n)$. This is a contradiction. Why?

1.5.4 It follows from Exercise 1.5.3 that $\sqrt{2}+1$ is irrational. Why does this imply that $\sqrt{2}$ is irrational?

1.6 Discussion

Euclid's *Elements* is the most influential book in the history of mathematics, and anyone interested in geometry should own a copy. It is not easy reading, but you will find yourself returning to it year after year and noticing something new. The standard edition in English is Heath's translation, which is now available as a Dover reprint of the 1925 Cambridge University Press edition. This reprint is carried by many bookstores; I have even seen it for sale at Los Angeles airport! Its main drawback is its size—three bulky volumes—due to the fact that more than half the content consists of

Heath's commentary. You can find the Heath translation *without* the commentary in the Britannica *Great Books of the Western World*, Volume 11. These books can often be found in used bookstores. Another, more recent, one-volume edition of the Heath translation is *Euclid's Elements*, edited by Dana Densmore and published by Green Lion Press in 2003.

A second (slight) drawback of the Heath edition is that it is about 80 years old and beginning to sound a little antiquated. Heath's English is sometimes quaint, and his commentary does not draw on modern research in geometry. He does not even mention some important advances that were known to experts in 1925. For this reason, a modern version of the *Elements* is desirable. A perfect version for the 21st century does not yet exist, but there is a nice concise web version by David Joyce at

`http://aleph0.clarkeu.edu/~djoyce/java/elements/elements.html`

This *Elements* has a small amount of commentary, but I mainly recommend it for proofs in simple modern English and nice diagrams. The diagrams are "variable" by dragging points on the screen, so each diagram represents all possible situations covered by a theorem.

For modern commentary on Euclid, I recommend two books: *Euclid: the Creation of Mathematics* by Benno Artmann and *Geometry: Euclid and Beyond* by Robin Hartshorne, published by Springer-Verlag in 1999 and 2000, respectively. Both books take Euclid as their starting point. Artmann mainly fills in the Greek background, although he also takes care to make it understandable to modern readers. Hartshorne is more concerned with what came after Euclid, and he gives a very thorough analysis of the gaps in Euclid and the ways they were filled by modern mathematicians. You will find Hartshorne useful supplementary reading for Chapters 2 and 3, where we examine the logical structure of the *Elements* and some of its gaps.

The climax of the *Elements* is the theory of regular polyhedra in Book XIII. Only five regular polyhedra exist, and they are shown in Figure 1.19. Notice that three of them are built from equilateral triangles, one from squares, and one from regular pentagons. This remarkable phenomenon underlines the importance of equilateral triangles and squares, and draws attention to the regular pentagon. In Chapter 2, we show how to construct it. Some geometers believe that the material in the *Elements* was chosen very much with the theory of regular polyhedra in mind. For example, Euclid wants to construct the equilateral triangle, square, and pentagon in order to construct the regular polyhedra.

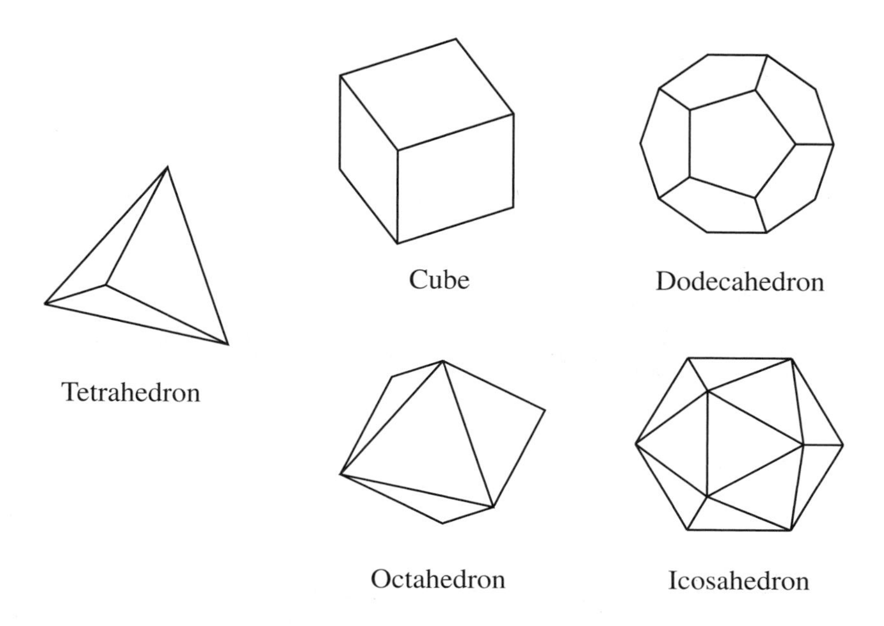

Figure 1.19: The regular polyhedra

It is fortunate that Euclid did not need regular polygons more complex than the pentagon, because none were constructed until modern times. The regular 17-gon was constructed by the 19-year-old Carl Friedrich Gauss in 1796, and his discovery was the key to the "question arising" from the construction of the equilateral triangle in Section 1.2: for which n is the regular n-gon constructible? Gauss showed (with some steps filled in by Pierre Wantzel in 1837) that a regular polygon with a prime number p of sides is constructible just in case p is of the form $2^{2^m}+1$. This result gives three constructible p-gons not known to the Greeks, because

$$2^4+1=17, \quad 2^8+1=257, \quad 2^{16}+1=65537$$

are all prime numbers. But no larger prime numbers of the form $2^{2^m}+1$ are known! Thus we do not know whether a larger constructible p-gon exists.

These results show that the *Elements* is not all of geometry, even if one accepts the same subject matter as Euclid. To see where Euclid fits in the general panorama of geometry, I recommend the books *Geometry and the Imagination* by D. Hilbert and S. Cohn-Vossen, and *Introduction to Geometry* by H. S. M. Coxeter (Wiley, 1969).

2

Euclid's approach to geometry

PREVIEW

Length is the fundamental concept of Euclid's geometry, but several important theorems seem to be "really" about angle or area—for example, the theorem on the sum of angles in a triangle and the Pythagorean theorem on the sum of squares. Also, Euclid often uses area to prove theorems about length, such as the Thales theorem.

In this chapter, we retrace some of Euclid's steps in the theory of angle and area to show how they lead to the Pythagorean theorem and the Thales theorem. We begin with his theory of angle, which shows most clearly the influence of his *parallel axiom*, the defining axiom of what is now called *Euclidean geometry*.

Angle is linked with length from the beginning by the so-called SAS ("side angle side") criterion for equal triangles (or "congruent triangles," as we now call them). We observe the implications of SAS for isosceles triangles and the properties of angles in a circle, and we note the related criterion, ASA ("angle side angle").

The theory of area depends on ASA, and it leads directly to a proof of the Pythagorean theorem. It leads more subtly to the Thales theorem and its consequences that we saw in Chapter 1. The theory of angle then combines nicely with the Thales theorem to give a second proof of the Pythagorean theorem.

In following these deductive threads, we learn more about the scope of straightedge and compass constructions, partly in the exercises. Interesting spinoffs from these investigations include a process for cutting any polygon into pieces that form a square, a construction for the square root of any length, and a construction of the regular pentagon.

2.1 The parallel axiom

In Chapter 1, we saw how useful it is to have *rectangles*: four-sided polygons whose angles are all right angles. Rectangles owe their existence to *parallel lines*—lines that do not meet—and fundamentally to the *parallel axiom* that Euclid stated as follows.

Euclid's parallel axiom. *If a straight line crossing two straight lines makes the interior angles on one side together less than two right angles, then the two straight lines will meet on that side.*

Figure 2.1 illustrates the situation described by Euclid's parallel axiom, which is what happens when the two lines are *not* parallel. If $\alpha+\beta$ is less than two right angles, then $\mathscr{L}$ and $\mathscr{M}$ meet somewhere on the right.

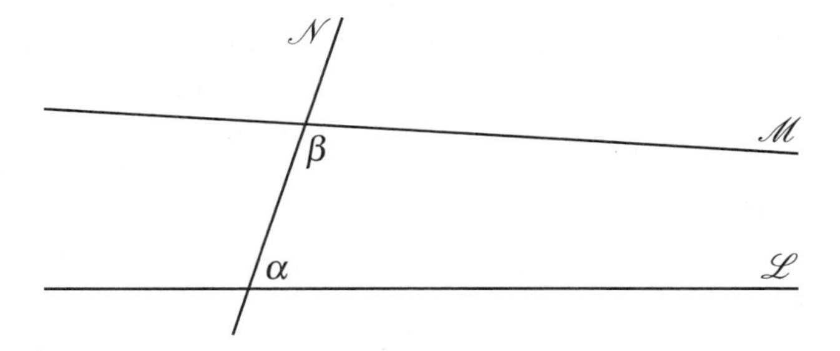

Figure 2.1: When lines are not parallel

It follows that if $\mathscr{L}$ and $\mathscr{M}$ do not meet on either side, then $\alpha+\beta=\pi$. In other words, if $\mathscr{L}$ and $\mathscr{M}$ *are* parallel, then α and β together make a straight angle and the angles made by $\mathscr{L}$, $\mathscr{M}$, and $\mathscr{N}$ are as shown in Figure 2.2.

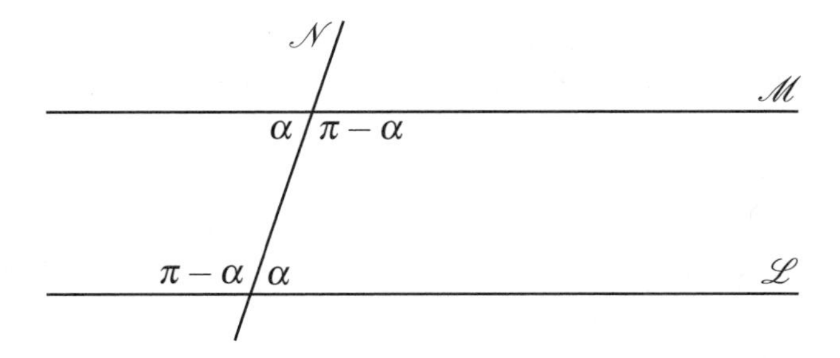

Figure 2.2: When lines are parallel

It also follows that any line through the intersection of $\mathscr{N}$ and $\mathscr{M}$, not meeting $\mathscr{L}$, makes the angle $\pi - \alpha$ with $\mathscr{N}$. Hence, this line equals $\mathscr{M}$. That is, *if a parallel to $\mathscr{L}$ through a given point exists, it is unique.*

It is a little more subtle to show the existence of a parallel to $\mathscr{L}$ through a given point P, but one way is to appeal to a principle called ASA ("angle side angle"), which will be discussed in Section 2.2.

Suppose that the lines $\mathscr{L}$, $\mathscr{M}$, and $\mathscr{N}$ make angles as shown in Figure 2.2, and that $\mathscr{L}$ and $\mathscr{M}$ are *not* parallel. Then, on at least one side of $\mathscr{N}$, there is a triangle whose sides are the segment of $\mathscr{N}$ between $\mathscr{L}$ and $\mathscr{M}$ and the segments of $\mathscr{L}$ and $\mathscr{M}$ between $\mathscr{N}$ and the point where they meet. According to ASA, this triangle is completely determined by the angles α, $\pi - \alpha$ and the segment of $\mathscr{N}$ between them. But then an identical triangle is determined on the other side of $\mathscr{N}$, and hence $\mathscr{L}$ and $\mathscr{M}$ also meet on the other side. This result contradicts Euclid's assumption (implicit in the construction axioms discussed in Section 1.1) that *there is a unique line through any two points*. Hence, the lines $\mathscr{L}$ and $\mathscr{M}$ are in fact parallel when the angles are as shown in Figure 2.2.

Thus, both the existence and the uniqueness of parallels follow from Euclid's parallel axiom (existence "follows trivially," because Euclid's parallel axiom is not required). It turns out that they also imply it, so the parallel axiom can be stated equivalently as follows.

Modern parallel axiom. *For any line $\mathscr{L}$ and point P outside $\mathscr{L}$, there is exactly one line through P that does not meet $\mathscr{L}$.*

This form of the parallel axiom is often called "Playfair's axiom," after the Scottish mathematician John Playfair who used it in a textbook in 1795. Playfair's axiom is simpler in form than Euclid's, because it does not involve angles, and this is often convenient. However, we often need parallel lines *and* the equal angles they create, the so-called *alternate interior angles* (for example, the angles marked α in Figure 2.2). In such situations, we prefer to use Euclid's parallel axiom.

Angles in a triangle

The existence of parallels and the equality of alternate interior angles imply a beautiful property of triangles.

Angle sum of a triangle. *If α, β, and γ are the angles of any triangle, then $\alpha + \beta + \gamma = \pi$.*

To prove this property, draw a line $\mathscr{L}$ through one vertex of the triangle, parallel to the opposite side, as shown in Figure 2.3.

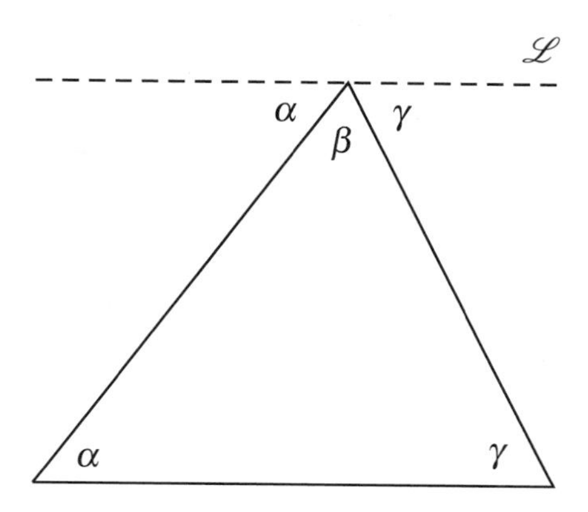

Figure 2.3: The angle sum of a triangle

Then the angle on the left beneath $\mathscr{L}$ is alternate to the angle α in the triangle, so it is equal to α. Similarly, the angle on the right beneath $\mathscr{L}$ is equal to γ. But then the straight angle π beneath $\mathscr{L}$ equals $\alpha+\beta+\gamma$, the angle sum of the triangle. □

Exercises

The triangle is the most important polygon, because any polygon can be built from triangles. For example, the angle sum of any quadrilateral (polygon with four sides) can be worked out by cutting the quadrilateral into two triangles.

2.1.1 Show that the angle sum of any quadrilateral is 2π.

A polygon $\mathscr{P}$ is called *convex* if the line segment between any two points in $\mathscr{P}$ lies entirely in $\mathscr{P}$. For these polygons, it is also easy to find the angle sum.

2.1.2 Explain why a convex n-gon can be cut into $n-2$ triangles.

2.1.3 Use the dissection of the n-gon into triangles to show that the angle sum of a convex n-gon is $(n-2)\pi$.

2.1.4 Use Exercise 2.1.3 to find the angle at each vertex of a *regular* n-gon (an n-gon with equal sides and equal angles).

2.1.5 Deduce from Exercise 2.1.4 that copies of a regular n-gon can tile the plane only for $n=3,4,6$.

2.2 Congruence axioms

Euclid says that two geometric figures *coincide* when one of them can be moved to fit exactly on the other. He uses the idea of moving one figure to coincide with another in the proof of Proposition 4 of Book I: *If two triangles have two corresponding sides equal, and the angles between these sides equal, then their third sides and the corresponding two angles are also equal.*

His proof consists of moving one triangle so that the equal angles of the two triangles coincide, and the equal sides as well. But then the third sides necessarily coincide, because their endpoints do, and hence, so do the other two angles.

Today we say that two triangles are *congruent* when their corresponding angles and side lengths are equal, and we no longer attempt to prove the proposition above. Instead, we *take it as an axiom* (that is, an unproved assumption), because it seems simpler to assume it than to introduce the concept of motion into geometry. The axiom is often called SAS (for "side angle side").

SAS axiom. *If triangles ABC and $A'B'C'$ are such that*

$$|AB| = |A'B'|, \quad \text{angle } ABC = \text{angle } A'B'C', \quad |BC| = |B'C'|$$

then also

$$|AC| = |A'C'|, \quad \text{angle } BCA = \text{angle } B'C'A', \quad \text{angle } CAB = \text{angle } C'A'B'.$$

For brevity, one often expresses SAS by saying that two triangles are congruent if two sides and the included angle are equal. There are similar conditions, ASA and SSS, which also imply congruence (but SSA does not—can you see why?). They can be deduced from SAS, so it is not necessary to take them as axioms. However, we will assume ASA here to save time, because it seems just as natural as SAS.

One of the most important consequences of SAS is Euclid's Proposition 5 of Book I. It says that a triangle with two equal sides has two equal angles. Such a triangle is called *isosceles*, from the Greek for "equal sides." The spectacular proof below is not from Euclid, but from the Greek mathematician Pappus, who lived around 300 CE.

Isosceles triangle theorem. *If a triangle has two equal sides, then the angles opposite to these sides are also equal.*

Suppose that triangle ABC has $|AB| = |AC|$. Then triangles ABC and ACB, which of course are the *same* triangle, are congruent by SAS (Figure 2.4). Their left sides are equal, their right sides are equal, and so are the angles between their left and right sides, because they are the *same* angle (the angle at A).

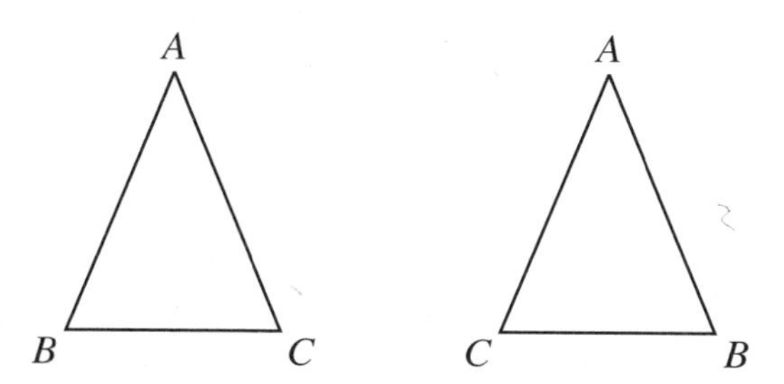

Figure 2.4: Two views of an isosceles triangle

But then it follows from SAS that all corresponding angles of these triangles are equal: for example, the bottom left angles. In other words, the angle at B equals the angle at C, so the angles opposite to the equal sides are equal. □

A useful consequence of ASA is the following theorem about parallelograms, which enables us to determine the area of triangles. (Remember, a parallelogram is defined as a figure bounded by two pairs of parallel lines—the definition does not say anything about the lengths of its sides.)

Parallelogram side theorem. *Opposite sides of a parallelogram are equal.*

To prove this theorem we divide the parallelogram into triangles by a diagonal (Figure 2.5), and try to prove that these triangles are congruent. They are, because

- they have the common side AC,
- their corresponding angles α are equal, being alternate interior angles for the parallels AD and BC,
- their corresponding angles β are equal, being alternate interior angles for the parallels AB and DC.

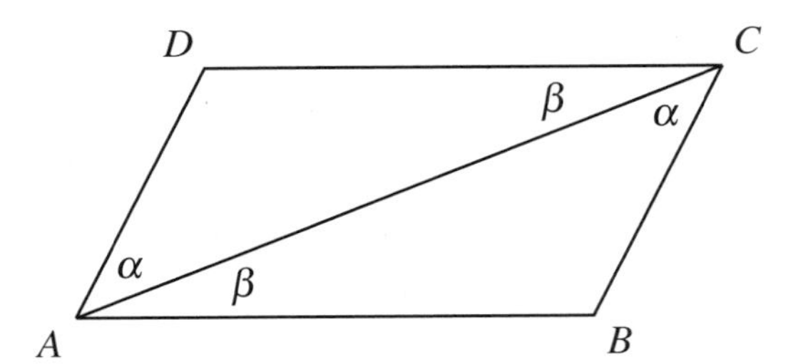

Figure 2.5: Dividing a parallelogram into triangles

Therefore, the triangles are congruent by ASA, and in particular we have the equalities $|AB| = |DC|$ and $|AD| = |BC|$ between corresponding sides. But these are also the opposite sides of the parallelogram. □

Exercises

2.2.1 Using the parallelogram side theorem and ASA, find congruent triangles in Figure 2.6. Hence, show that the diagonals of a parallelogram bisect each other.

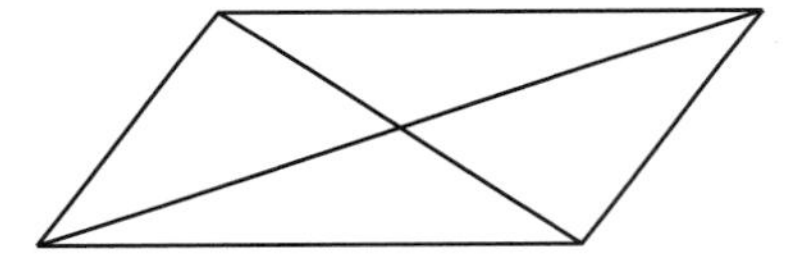

Figure 2.6: A parallelogram and its diagonals

2.2.2 Deduce that the diagonals of a rhombus—a parallelogram whose sides are all equal—meet at right angles. (*Hint*: You may find it convenient to use SSS, which says that triangles are congruent when their corresponding sides are equal.)

2.2.3 Prove the isosceles triangle theorem differently by bisecting the angle at A.

2.3 Area and equality

The principle of logic used in Section 1.2—that things equal to the same thing are equal to each other—is one of five principles that Euclid calls *common notions*. The common notions he states are particularly important for his theory of area, and they are as follows:

1. Things equal to the same thing are also equal to one another.
2. If equals are added to equals, the wholes are equal.
3. If equals are subtracted from equals, the remainders are equal.
4. Things that coincide with one another are equal to one another.
5. The whole is greater than the part.

The word "equal" here means "equal in some specific respect." In most cases, it means "equal in length" or "equal in area," although Euclid's idea of "equal in area" is not exactly the same as ours, as I will explain below. Likewise, "addition" can mean addition of lengths or addition of areas, but Euclid never adds a length to an area because this has no meaning in his system.

A simple but important example that illustrates the use of "equals" is Euclid's Proposition 15 of Book I: *Vertically opposite angles are equal.* Vertically opposite angles are the angles α shown in Figure 2.7.

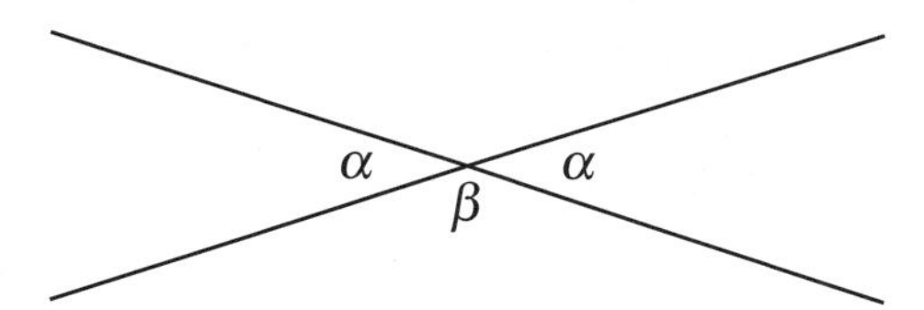

Figure 2.7: Vertically opposite angles

They are equal because each of them equals a straight angle minus β.

The square of a sum

Proposition 4 of Book II is another interesting example. It states a property of squares and rectangles that we express by the algebraic formula

$$(a+b)^2 = a^2 + 2ab + b^2.$$

Euclid does *not* have algebraic notation, so he has to state this equation in words: *If a line is cut at random, the square on the whole is equal to the squares on the segments and twice the rectangle contained by the segments.* Whichever way you say it, Figure 2.8 explains why it is true.

The line is $a+b$ because it is cut into the two segments a and b, and hence

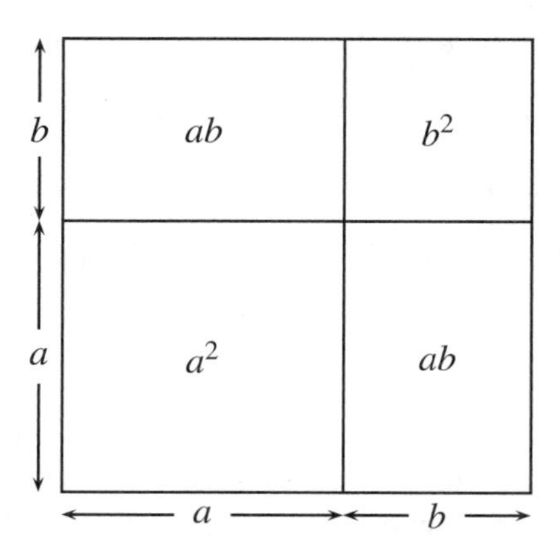

Figure 2.8: The square of a sum of line segments

- The square on the line is what we write as $(a+b)^2$.
- The squares on the two segments a and b are a^2 and b^2, respectively.
- The rectangle "contained" by the segments a and b is ab.
- The square $(a+b)^2$ equals (in area) the sum of a^2, b^2, and two copies of ab.

It should be emphasized that, in Greek mathematics, the *only* interpretation of ab, the "product" of line segments a and b, is the rectangle with perpendicular sides a and b (or "contained in" a and b, as Euclid used to say). This rectangle could be shown "equal" to certain other regions, but only by cutting the regions into identical pieces by straight lines. The Greeks did not realize that this "equality of regions" was the same as equality of numbers—the numbers we call the *areas* of the regions—partly because they did not regard irrational lengths as numbers, and partly because they did not think the product of lengths should be a length.

As mentioned in Section 1.5, this belief was not necessarily an obstacle to the development of geometry. To find the area of nonrectangular regions, such as triangles or parallelograms, one has to think about cutting regions into pieces in any case. For such simple regions, there is no particular advantage in thinking of the area as a number, as we will see in Section 2.4. But first we need to investigate the concept mentioned in Euclid's Common Notion number 4. What does it mean for one figure to "coincide" with another?

Exercises

In Figure 2.8, the large square is subdivided by two lines: one of them perpendicular to the bottom side of the square and the other perpendicular to the left side of the square.

2.3.1 Use the parallel axiom to explain why all other angles in the figure are necessarily right angles.

Figure 2.8 presents the algebraic identity $(a+b)^2 = a^2 + 2ab + b^2$ in geometric form. Other well-known algebraic identities can also be given a geometric presentation.

2.3.2 Give a diagram for the identity $a(b+c) = ab+ac$.

2.3.3 Give a diagram for the identity $a^2 - b^2 = (a+b)(a-b)$.

Euclid does not give a geometric theorem that explains the identity $(a+b)^3 = a^3 + 3a^2b + 3ab^2 + b^3$. But it is not hard to do so by interpreting $(a+b)^3$ as a cube with edge length $a+b$, a^3 as a cube with edge a, a^2b as a box with perpendicular edges a, a, and b, and so on.

2.3.4 Draw a picture of a cube with edges $a+b$, and show it cut by planes (parallel to its faces) that divide each edge into a segment of length a and a segment of length b.

2.3.5 Explain why these planes cut the original cube into eight pieces:

- a cube with edges a,
- a cube with edges b,
- three boxes with edges a, a, b,
- three boxes with edges a, b, b.

2.4 Area of parallelograms and triangles

The first nonrectangular region that can be shown "equal" to a rectangle in Euclid's sense is a parallelogram. Figure 2.9 shows how to use straight lines to cut a parallelogram into pieces that can be reassembled to form a rectangle.

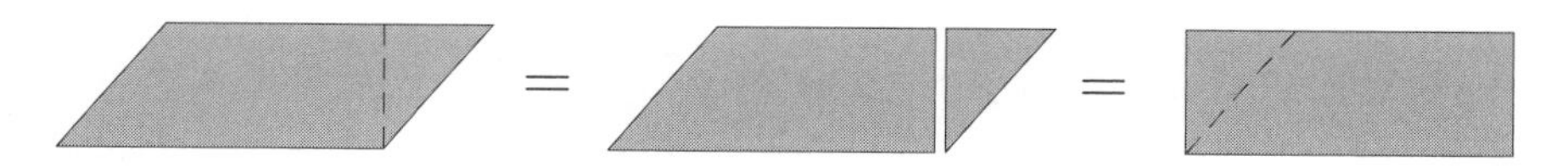

Figure 2.9: Assembling parallelogram and rectangle from the same pieces

Only one cut is needed in the example of Figure 2.9, but more cuts are needed if the parallelogram is more sheared, as in Figure 2.10.

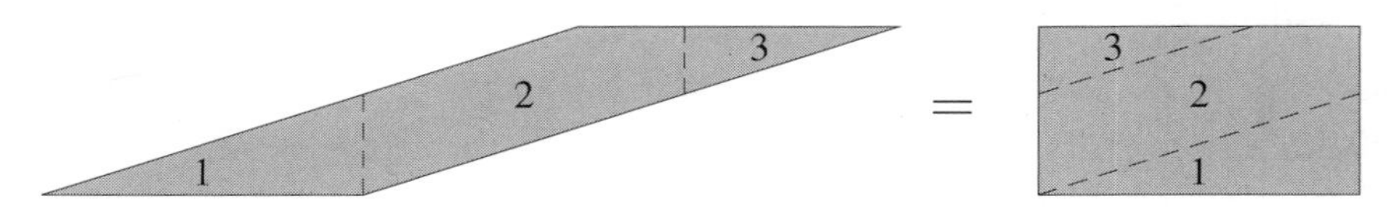

Figure 2.10: A case in which more cuts are required

In Figure 2.10 we need two cuts, which produce the pieces labeled 1, 2, 3. The number of cuts can become arbitrarily large as the parallelogram is sheared further. We can avoid large numbers of cuts by allowing *subtraction* of pieces as well as addition. Figure 2.11 shows how to convert any rectangle to any parallelogram with the same *base* OR and the same *height* OP. We need only add a triangle, and then subtract an equal triangle.

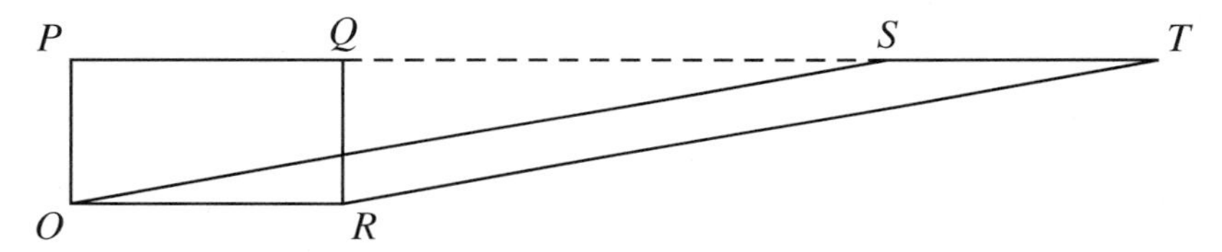

Figure 2.11: Rectangle and parallelogram with the same base and height

To be precise, if we start with rectangle $OPQR$ and add triangle RQT, then subtract triangle OPS (which equals triangle RQT by the parallelogram side theorem of Section 2.2), the result is parallelogram $OSTR$. Thus, the parallelogram is equal (in area) to a rectangle with the same base and height. We write this fact as

$$\text{area of parallelogram} = \text{base} \times \text{height}.$$

To find the area of a triangle ABC, we notice that it can be viewed as "half" of a parallelogram by adding to it the congruent triangle ACD as shown in Figure 2.5, and again in Figure 2.12.

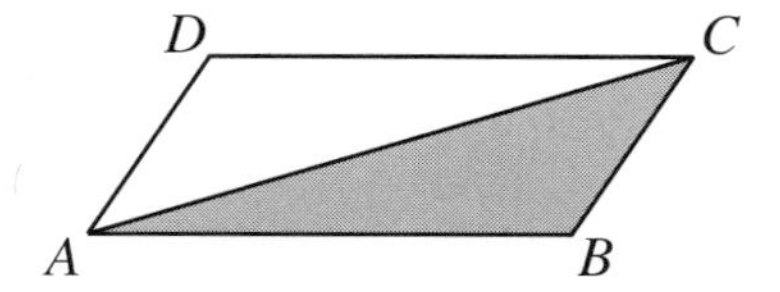

Figure 2.12: A triangle as half a parallelogram

Clearly,

$$\text{area of triangle } ABC + \text{area of triangle } ACD = \text{area of parallelogram } ABCD,$$

and the two triangles "coincide" (because they are congruent) and so they have equal area by Euclid's Common Notion 4. Thus,

$$\text{area of triangle} = \frac{1}{2}\,\text{base} \times \text{height}.$$

This formula is important in two ways:

- *As a statement about area.* From a modern viewpoint, the formula gives the area of the triangle as a product of numbers. From the ancient viewpoint, it gives a rectangle "equal" to the triangle, namely, the rectangle with the same base and half the height of the triangle.
- *As a statement about proportionality.* For triangles with the same height, the formula shows that their areas are proportional to their bases. This statement turns out to be crucial for the proof of the Thales theorem (Section 2.6).

The proportionality statement follows from the assumption that each line segment has a real number length, which depends on the acceptance of irrational numbers. As mentioned in the previous section, the Greeks did not accept this assumption. Euclid got the proportionality statement by a lengthy and subtle "theory of proportion" in Book V of the *Elements*.

Exercises

To back up the claim that the formula $\frac{1}{2}$ base $\times$ height gives a way to find the area of the triangle, we should explain how to find the height.

2.4.1 Given a triangle with a particular side specified as the "base," show how to find the height by straightedge and compass construction.

The equality of triangles OPS and RQT follows from the parallelogram side theorem, as claimed above, but a careful proof would explain what other axioms are involved.

2.4.2 By what Common Notion does $|PQ| = |ST|$?

2.4.3 By what Common Notion does $|PS| = |QT|$?

2.4.4 By what congruence axiom is triangle OPS congruent to triangle RQT?

2.5 The Pythagorean theorem

The Pythagorean theorem is about areas, and indeed Euclid proves it immediately after he has developed the theory of area for parallelograms and triangles in Book I of the *Elements*. First let us recall the statement of the theorem.

Pythagorean theorem. *For any right-angled triangle, the sum of the squares on the two shorter sides equals the square on the hypotenuse.*

We follow Euclid's proof, in which he divides the square on the hypotenuse into the two rectangles shown in Figure 2.13. He then shows that the light gray square equals the light gray rectangle and that the dark gray square equals the dark gray rectangle, so the sum of the light and dark squares is the square on the hypotenuse, as required.

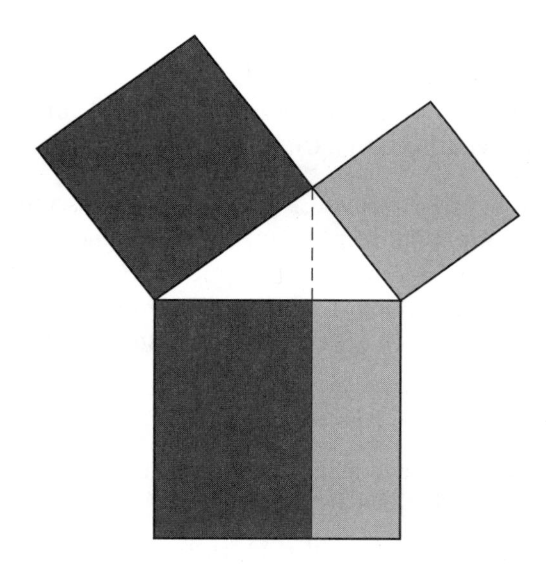

Figure 2.13: Dividing the square for Euclid's proof

First we show equality for the light gray regions in Figure 2.13, and in fact we show that *half* of the light gray square equals half of the light gray rectangle. We start with a light gray triangle that is obviously half of the light gray square, and we successively replace it with triangles of the same base or height, ending with a triangle that is obviously half of the light gray rectangle (Figure 2.14).

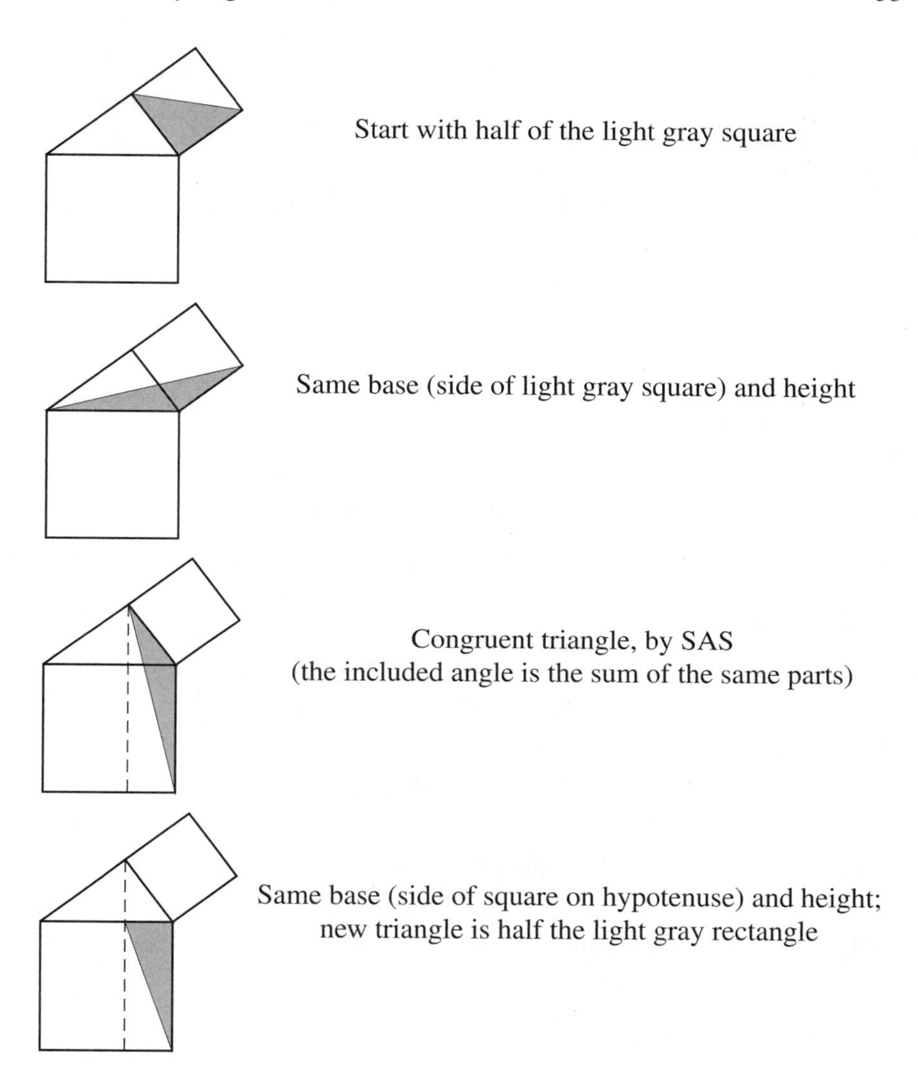

Figure 2.14: Changing the triangle without changing its area

The same argument applies to the dark gray regions, and thus, the Pythagorean theorem is proved. □

Figure 2.13 suggests a natural way to construct a square equal in area to a given rectangle. Given the light gray rectangle, say, the problem is to reconstruct the rest of Figure 2.13.

We can certainly extend a given rectangle to a square and hence reconstruct the square on the hypotenuse. The main problem is to reconstruct the right-angled triangle, from the hypotenuse, so that the other vertex lies on the dashed line. See whether you can think of a way to do this; a really elegant solution is given in Section 2.7. Once we have the right-angled triangle, we can certainly construct the squares on its other two sides—in particular, the gray square equal in area to the gray rectangle.

Exercises

It follows from the Pythagorean theorem that a right-angled triangle with sides 3 and 4 has hypotenuse $\sqrt{3^2+4^2} = \sqrt{25} = 5$. But there is *only one* triangle with sides 3, 4, and 5 (by the SSS criterion mentioned in Exercise 2.2.2), so putting together lengths 3, 4, and 5 always makes a right-angled triangle. This triangle is known as the $(3,4,5)$ triangle.

2.5.1 Verify that the $(5,12,13)$, $(8,15,17)$, and $(7,24,25)$ triangles are right-angled.

2.5.2 Prove the converse Pythagorean theorem: If $a,b,c > 0$ and $a^2+b^2 = c^2$, then the triangle with sides a,b,c is right-angled.

2.5.3 How can we be sure that lengths $a,b,c > 0$ with $a^2+b^2 = c^2$ actually fit together to make a triangle? (*Hint:* Show that $a+b > c$.)

Right-angled triangles can be used to construct certain irrational lengths. For example, we saw in Section 1.5 that the right-angled triangle with sides 1, 1 has hypotenuse $\sqrt{2}$.

2.5.4 Starting from the triangle with sides 1, 1, and $\sqrt{2}$, find a straightedge and compass construction of $\sqrt{3}$.

2.5.5 Hence, obtain constructions of $\sqrt{n}$ for $n = 2,3,4,5,6,\ldots$.

2.6 Proof of the Thales theorem

We mentioned this theorem in Chapter 1 as a fact with many interesting consequences, such as the proportionality of similar triangles. We are now in a position to prove the theorem as Euclid did in his Proposition 2 of Book VI. Here again is a statement of the theorem.

The Thales theorem. *A line drawn parallel to one side of a triangle cuts the other two sides proportionally.*

The proof begins by considering triangle ABC, with its sides AB and AC cut by the parallel PQ to side BC (Figure 2.15). Because PQ is parallel to BC, the triangles PQB and PQC on base PQ have the same height, namely the distance between the parallels. They therefore have the same area.

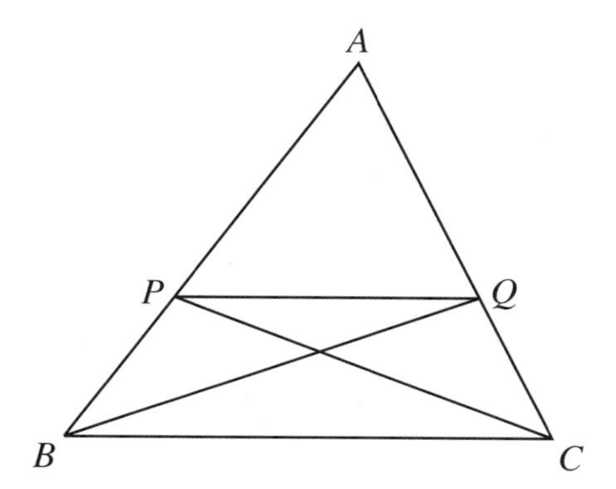

Figure 2.15: Triangle sides cut by a parallel

If we add triangle APQ to each of the equal-area triangles PQB and PQC, we get the triangles AQB and APC, respectively. Hence, the latter triangles are also equal in area.

Now consider the two triangles—APQ and PQB—that make up triangle AQB as triangles with bases on the line AB. They have the same height relative to this base (namely, the perpendicular distance of Q from AB). Hence, their bases are in the ratio of their areas:

$$\frac{|AP|}{|PB|} = \frac{\text{area } APQ}{\text{area } PQB}.$$

Similarly, considering the triangles APQ and PQC that make up the triangle APC, we find that

$$\frac{|AQ|}{|QC|} = \frac{\text{area } APQ}{\text{area } PQC}.$$

Because area PQB equals area PQC, the right sides of these two equations are equal, and so are their left sides. That is,

$$\frac{|AP|}{|PB|} = \frac{|AQ|}{|QC|}.$$

In other words, the line PQ cuts the sides AB and AC proportionally. □

Exercises

As seen in Exercise 1.3.6, $|AP|/|PB| = |AQ|/|QC|$ is equivalent to $|AP|/|AB| = |AQ|/|AC|$. This equation is a more convenient formulation of the Thales theorem if you want to prove the following generalization:

2.6.1 Suppose that there are several parallels $P_1Q_1, P_2Q_2, P_3Q_3 \ldots$ to the side BC of triangle ABC. Show that

$$\frac{|AP_1|}{|AQ_1|} = \frac{|AP_2|}{|AQ_2|} = \frac{|AP_3|}{|AQ_3|} = \cdots = \frac{|AB|}{|AC|}.$$

We can also drop the assumption that the parallels $P_1Q_1, P_2Q_2, P_3Q_3 \ldots$ fall across a triangle ABC.

2.6.2 If parallels $P_1Q_1, P_2Q_2, P_3Q_3 \ldots$ fall across a pair of *parallel* lines $\mathscr{L}$ and $\mathscr{M}$, what can we say about the lengths they cut from $\mathscr{L}$ and $\mathscr{M}$?

2.7 Angles in a circle

The isosceles triangle theorem of Section 2.2, simple though it is, has a remarkable consequence.

Invariance of angles in a circle. *If A and B are two points on a circle, then, for all points C on one of the arcs connecting them, the angle ACB is constant.*

To prove invariance we draw lines from A, B, C to the center of the circle, O, along with the lines making the angle ACB (Figure 2.16).

Because all radii of the circle are equal, $|OA| = |OC|$. Thus triangle AOC is isosceles, and the angles α in it are equal by the isosceles triangle theorem. The angles β in triangle BOC are equal for the same reason.

Because the angle sum of any triangle is π (Section 2.1), it follows that the angle at O in triangle AOC is $\pi - 2\alpha$ and the angle at O in triangle BOC is $\pi - 2\beta$. It follows that the third angle at O, angle AOB, is $2(\alpha + \beta)$, because the total angle around any point is 2π. But angle AOB is *constant*, so $\alpha + \beta$ is also constant, and $\alpha + \beta$ is precisely the angle at C. □

An important special case of this theorem is when A, O, and B lie in a straight line, so $2(\alpha + \beta) = \pi$. In this case, $\alpha + \beta = \pi/2$, and thus we have the following theorem (which is also attributed to Thales).

Angle in a semicircle theorem. *If A and B are the ends of a diameter of a circle, and C is any other point on the circle, then angle ACB is a right angle.* □

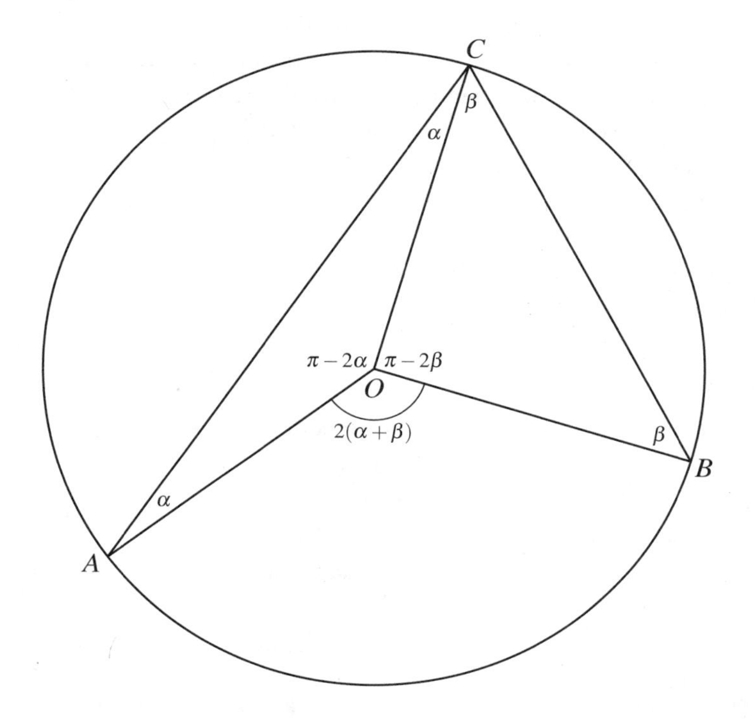

Figure 2.16: Angle $\alpha + \beta$ in a circle

This theorem enables us to solve the problem left open at the end of Section 2.5: Given a hypotenuse AB, how do we construct the right-angled triangle whose other vertex C lies on a given line? Figure 2.17 shows how.

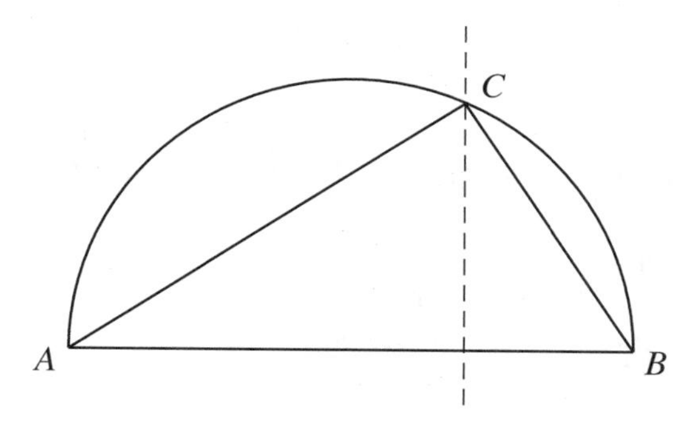

Figure 2.17: Constructing a right-angled triangle with given hypotenuse

The trick is to draw the semicircle on diameter AB, which can be done by first bisecting AB to obtain the center of the circle. Then the point where the semicircle meets the given line (shown dashed) is necessarily the other vertex C, because the angle at C is a right angle.

This construction completes the solution of the problem raised at the end of Section 2.5: finding a square equal in area to a given rectangle. In Section 2.8 we will show that Figure 2.17 also enables us to construct the *square root* of an arbitrary length, and it gives a new proof of the Pythagorean theorem.

Exercises

2.7.1 Explain how the angle in a semicircle theorem enables us to construct a right-angled triangle with a given hypotenuse AB.

2.7.2 Then, by looking at Figure 2.13 from the bottom up, find a way to construct a square equal in area to a given rectangle.

2.7.3 Given any two squares, we can construct a square that equals (in area) the sum of the two given squares. Why?

2.7.4 Deduce from the previous exercises that any polygon may be "squared"; that is, there is a straightedge and compass construction of a square equal in area to the given polygon. (You may assume that the given polygon can be cut into triangles.)

The possibility of "squaring" any polygon was apparently known to Greek mathematicians, and this may be what tempted them to try "squaring the circle": constructing a square equal in area to a given circle. There is no straightedge and compass solution of the latter problem, but this was not known until 1882.

Coming back to angles in the circle, here is another theorem about invariance of angles:

2.7.5 If a quadrilateral has its vertices on a circle, show that its opposite angles sum to π.

2.8 The Pythagorean theorem revisited

In Book VI, Proposition 31 of the *Elements*, Euclid proves a generalization of the Pythagorean theorem. From it, we get a new proof of the ordinary Pythagorean theorem, based on the proportionality of similar triangles.

Given a right-angled triangle with sides a, b, and hypotenuse c, we divide it into two smaller right-angled triangles by the perpendicular to the hypotenuse through the opposite vertex (the dashed line in Figure 2.18).

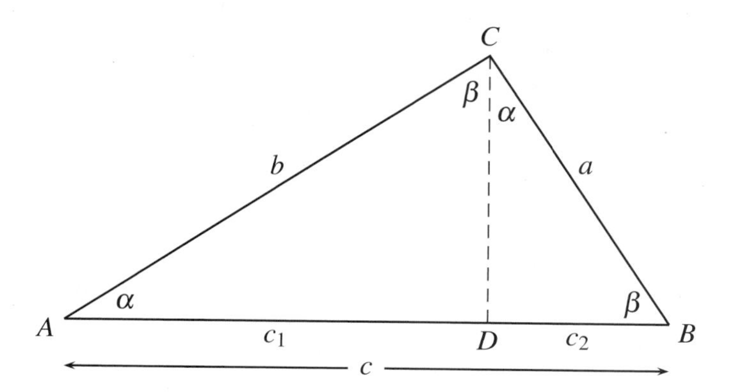

Figure 2.18: Subdividing a right-angled triangle into similar triangles

All three triangles are similar because they have the same angles α and β. If we look first at the angle α at A and the angle β at B, then

$$\alpha + \beta = \frac{\pi}{2}$$

because the angle sum of triangle ABC is π and the angle at C is $\pi/2$. But then it follows that angle $ACD = \beta$ in triangle ACD (to make its angle sum $= \pi$) and angle $DCB = \alpha$ in triangle DCB (to make its angle sum $= \pi$).

Now we use the proportionality of these triangles, calling the side opposite α in each triangle "short" and the side opposite β "long" for convenience. Comparing triangle ABC with triangle ADC, we get

$$\frac{\text{long side}}{\text{hypotenuse}} = \frac{b}{c} = \frac{c_1}{b}, \quad \text{hence} \quad b^2 = cc_1.$$

Comparing triangle ABC with triangle DCB, we get

$$\frac{\text{short side}}{\text{hypotenuse}} = \frac{a}{c} = \frac{c_2}{a}, \quad \text{hence} \quad a^2 = cc_2.$$

Adding the values of a^2 and b^2 just obtained, we finally get

$$a^2 + b^2 = cc_2 + cc_1 = c(c_1 + c_2) = c^2 \quad \text{because } c_1 + c_2 = c,$$

and this is the Pythagorean theorem. □

This second proof is not really shorter than Euclid's first (given in Section 2.5) when one takes into account the work needed to prove the proportionality of similar triangles. However, we often need similar triangles, so they are a standard tool, and a proof that uses standard tools is generally preferable to one that uses special machinery. Moreover, the splitting of a right-angled triangle into similar triangles is itself a useful tool—it enables us to construct the square root of any line segment.

Straightedge and compass construction of square roots

Given any line segment l, construct the semicircle with diameter $l+1$, and the perpendicular to the diameter where the segments 1 and l meet (Figure 2.19). Then *the length h of this perpendicular is* $\sqrt{l}$.

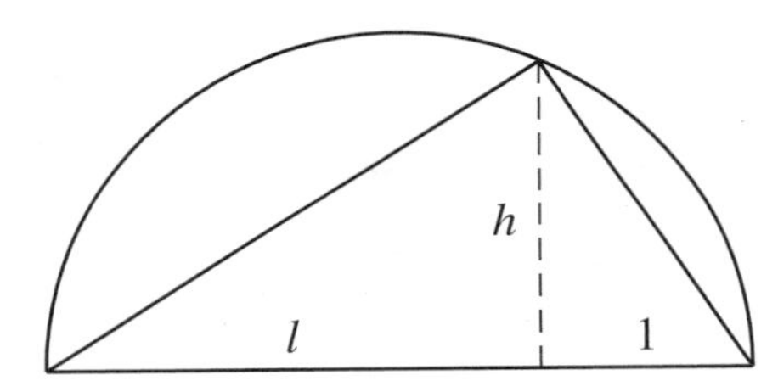

Figure 2.19: Construction of the square root

To see why, construct the right-angled triangle with hypotenuse $l+1$ and third vertex where the perpendicular meets the semicircle. We know that the perpendicular splits this triangle into two similar, and hence proportional, triangles. In the triangle on the left,

$$\frac{\text{long side}}{\text{short side}} = \frac{l}{h}.$$

In the triangle on the right,

$$\frac{\text{long side}}{\text{short side}} = \frac{h}{1}.$$

Because these ratios are equal by proportionality of the triangles, we have

$$\frac{l}{h} = \frac{h}{1},$$

hence $h^2 = l$; that is, $h = \sqrt{l}$. □

This result complements the constructions for the rational operations $+, -, \times$, and $\div$ we gave in Chapter 1. The constructibility of these and $\sqrt{}$ was first pointed out by Descartes in his book *Géométrie* of 1637. Rational operations and $\sqrt{}$ are in fact *precisely* what can be done with straightedge and compass. When we introduce coordinates in Chapter 3 we will see that any "constructible point" has coordinates obtainable from the unit length 1 by $+, -, \times, \div$, and $\sqrt{}$.

Exercises

Now that we know how to construct the $+, -, \times, \div$, and $\sqrt{}$ of given lengths, we can use algebra as a shortcut to decide whether certain figures are constructible by straightedge and compass. If we know that a certain figure is constructible from the length $(1+\sqrt{5})/2$, for example, then we know that the figure is constructible—period—because the length $(1+\sqrt{5})/2$ is built from the unit length by the operations $+, \times, \div$, and $\sqrt{}$.

This is precisely the case for the regular pentagon, which was constructed by Euclid in Book IV, Proposition 11, using virtually all of the geometry he had developed up to that point. We also need nearly everything we have developed up to this point, but it fills less space than four books of the *Elements*!

The following exercises refer to the regular pentagon of side 1 shown in Figure 2.20 and its diagonals of length x.

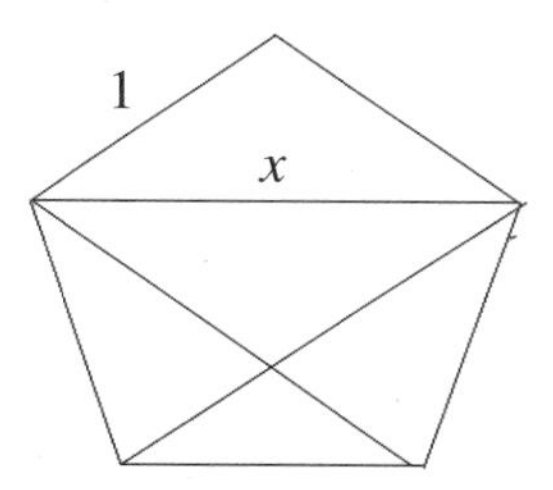

Figure 2.20: The regular pentagon

2.8.1 Use the symmetry of the regular pentagon to find similar triangles implying

$$\frac{x}{1} = \frac{1}{x-1},$$

that is, $x^2 - x - 1 = 0$.

2.8.2 By finding the positive root of this quadratic equation, show that each diagonal has length $x = (1+\sqrt{5})/2$.

2.8.3 Now show that the regular pentagon is constructible.

2.9 Discussion

Euclid found the most important axiom of geometry—the parallel axiom—and he also identified the basic theorems and traced the logical connections between them. However, his approach misses certain fine points and is not logically complete. For example, in his very first proof (the construction of the equilateral triangle), he assumes that certain circles have a point in common, but none of his axioms guarantee the existence of such a point. There are many such situations, in which Euclid assumes something is true because it *looks* true in the diagram.

Euclid's theory of area is a whole section of his geometry that seems to have no geometric support. Its concepts seem more like arithmetic—addition, subtraction, and proportion—but its concept of multiplication is not the usual one, because multiplication of more than three lengths is not allowed.

These gaps in Euclid's approach to geometry were first noticed in the 19th century, and the task of filling them was completed by David Hilbert in his *Grundlagen der Geometrie* (Foundations of Geometry) of 1899. On the one hand, Hilbert introduced axioms of *incidence* and *order*, giving the conditions under which lines (and circles) meet. These justify the belief that "geometric objects behave as the pictures suggest." On the other hand, Hilbert replaced Euclid's theory of area with a genuine arithmetic, which he called *segment arithmetic*. He defined the sum and product of segments as we did in Section 1.4 and proved that these operations on segments have the same properties as ordinary sum and product. For example,

$$a+b=b+a, \quad ab=ba, \quad a(b+c)=ab+ac, \quad \text{and so on.}$$

In the process, Hilbert discovered that the Pappus and Desargues theorems (Exercises 1.4.3 and 1.4.4) play a decisive role.

The downside of Hilbert's completion of Euclid is that it is lengthy and difficult. Nearly 20 axioms are required, and some key theorems are hard to prove. To some extent, this hardship occurs because Hilbert insists on geometric definitions of $+$ and $\times$. He wants numbers to come from "inside" geometry rather than from "outside". Thus, to prove that $ab=ba$ he needs the theorem of Pappus, and to prove that $a(bc)=(ab)c$ he needs the theorem of Desargues.

Even today, the construction of segment arithmetic is an admirable feat. As Hilbert pointed out, it shows that Euclid was right to believe that the theory of proportion could be developed without new geometric axioms. Still, it is somewhat quixotic to build numbers "inside" Euclid's geometry when they are brought from "outside" into nearly every other branch of geometry. It is generally easier to build geometry on numbers than the reverse, and Euclidean geometry is no exception, as I hope to show in Chapters 3 and 4.

This is one reason for bypassing Hilbert's approach, so I will merely list his axioms here. They are thoroughly investigated in Hartshorne's *Geometry: Euclid and Beyond* or Hilbert's own book, which is available in English translation. Hartshorne's book has the clearest available derivation of ordinary geometry and segment arithmetic from the Hilbert axioms, so it should be consulted by anyone who wants to see Euclid's approach taken to its logical conclusion.

There is another reason to bypass Hilbert's axioms, apart from their difficulty. In my opinion, Hilbert's greatest geometric achievement was to build arithmetic, not in *Euclidean* geometry, but in *projective* geometry. As just mentioned, Hilbert found that the keys to segment arithmetic are the Pappus and Desargues theorems. These two theorems do not involve the concept of length, and so they really belong to a more primitive kind of geometry. This primitive geometry (projective geometry) has only a handful of axioms—*fewer than the usual axioms for arithmetic*—so it is more interesting to build arithmetic inside it. It is also less trouble, because we do not have to *prove* the Pappus and Desargues theorems. We will explain how projective geometry contains arithmetic in Chapters 5 and 6.

Hilbert's axioms

The axioms concern undefined objects called "points" and "lines," the related concepts of "line segment," "ray," and "angle," and the relations of "betweenness" and "congruence." Following Hartshorne, we simplify Hilbert's axioms slightly by stating some of them in a stronger form than necessary.

The first group of axioms is about *incidence*: conditions for points to lie on lines or for lines to pass through points.

I1. For any two points A, B, a unique line passes through A, B.

I2. Every line contains at least two points.

I3. There exist three points not all on the same line.

I4. For each line $\mathscr{L}$ and point P not on $\mathscr{L}$ there is a unique line through P not meeting $\mathscr{L}$ (parallel axiom).

The next group is about *betweenness* or *order*: a concept overlooked by Euclid, probably because it is too "obvious." The first to draw attention to betweenness was the German mathematician Moritz Pasch, in the 1880s. We write $A * B * C$ to denote that B is between A and C.

B1. If $A * B * C$, then A, B, C are three points on a line and $C * B * A$.

B2. For any two points A and B, there is a point C with $A * B * C$.

B3. Of three points on a line, exactly one is between the other two.

B4. Suppose A, B, C are three points not in a line and that $\mathscr{L}$ is a line not passing through any of A, B, C. If $\mathscr{L}$ contains a point D between A and B, then $\mathscr{L}$ contains either a point between A and C or a point between B and C, but not both (Pasch's axiom).

The next group is about *congruence of line segments* and *congruence of angles*, both denoted by $\cong$. Thus, $AB \cong CD$ means that AB and CD have equal length and $\angle ABC \cong \angle DEF$ means that $\angle ABC$ and $\angle DEF$ are equal angles. Notice that C2 and C5 contain versions of Euclid's Common Notion 1: "Things equal to the same thing are equal to each other."

C1. For any line segment AB, and any ray $\mathscr{R}$ originating at a point C, there is a unique point D on $\mathscr{R}$ with $AB \cong CD$.

C2. If $AB \cong CD$ and $AB \cong EF$, then $CD \cong EF$. For any AB, $AB \cong AB$.

C3. Suppose $A * B * C$ and $D * E * F$. If $AB \cong DE$ and $BC \cong EF$, then $AC \cong DF$. (Addition of lengths is well-defined.)

C4. For any angle $\angle BAC$, and any ray $\overrightarrow{DF}$, there is a unique ray $\overrightarrow{DE}$ on a given side of $\overrightarrow{DF}$ with $\angle BAC \cong \angle EDF$.

C5. For any angles α, β, γ, if $\alpha \cong \beta$ and $\alpha \cong \gamma$, then $\beta \cong \gamma$. Also, $\alpha \cong \alpha$.

C6. Suppose that ABC and DEF are triangles with $AB \cong DE$, $AC \cong DF$, and $\angle BAC \cong \angle EDF$. Then, the two triangles are congruent, namely $BC \cong EF$, $\angle ABC \cong \angle DEF$, and $\angle ACB \cong \angle DFE$. (This is SAS.)

Then there is an axiom about the intersection of circles. It involves the concept of points *inside* the circle, which are those points whose distance from the center is less than the radius.

E. Two circles meet if one of them contains points both inside and outside the other.

Next there is the so-called *Archimedean axiom*, which says that no length can be "infinitely large" relative to another.

A. For any line segments AB and CD, there is a natural number n such that n copies of AB are together greater than CD.

Finally, there is the so-called *Dedekind axiom*, which says that the line is *complete*, or has *no gaps*. It implies that its points correspond to real numbers. Hilbert wanted an axiom like this to force the plane of Euclidean geometry to be the same as the plane $\mathbb{R}^2$ of pairs of real numbers.

D. Suppose the points of a line $\mathscr{L}$ are divided into two nonempty subsets $\mathscr{A}$ and $\mathscr{B}$ in such a way that no point of $\mathscr{A}$ is between two points of $\mathscr{B}$ and no point of $\mathscr{B}$ is between two points of $\mathscr{A}$. Then, a unique point P, either in $\mathscr{A}$ or $\mathscr{B}$, lies between any other two points, of which one is in $\mathscr{A}$ and the other is in $\mathscr{B}$.

Axiom D is not needed to derive any of Euclid's theorems. They do not involve all real numbers but only the so-called *constructible* numbers originating from straightedge and compass constructions. However, who can be sure that we will never need nonconstructible points? One of the most important numbers in geometry, π, is nonconstructible! (Because the circle cannot be squared.) Thus, it seems prudent to use Axiom D so that the line is complete from the beginning.

In Chapter 3, we will take the real numbers as the starting point of geometry, and see what advantages this may have over the Euclid–Hilbert approach. One clear advantage is *access to algebra*, which reduces many geometric problems to simple calculations. Algebra also offers some conceptual advantages, as we will see.

3

Coordinates

PREVIEW

Around 1630, Pierre de Fermat and René Descartes independently discovered the advantages of numbers in geometry, as *coordinates*. Descartes was the first to publish a detailed account, in his book *Géométrie* of 1637. For this reason, he gets most of the credit for the idea and the coordinate approach to geometry became known as *Cartesian* (from the old way of writing his name: Des Cartes).

Descartes thought that geometry was as Euclid described it, and that numbers merely *assist* in studying geometric figures. But later mathematicians discovered objects with "non-Euclidean" properties, such as "lines" having more than one "parallel" through a given point. To clarify this situation, it became desirable to *define* points, lines, length, and so on, and to *prove* that they satisfy Euclid's axioms.

This program, carried out with the help of coordinates, is called the *arithmetization of geometry*. In the first three sections of this chapter, we do the main steps, using the set $\mathbb{R}$ of real numbers to define the *Euclidean plane* $\mathbb{R}^2$ and the points, lines, and circles in it. We also define the concepts of distance and (briefly) angle, and show how some crucial axioms and theorems follow. However, arithmetization does much more.

- It gives an algebraic description of constructibility by straightedge and compass (Section 3.4), which makes it possible to prove that certain figures are *not* constructible.
- It enables us to define what it means to "move" a geometric figure (Section 3.6), which provides justification for Euclid's proof of SAS, and raises a new kind of geometric question (Section 3.7): What kinds of "motion" exist?

3.1 The number line and the number plane

The set $\mathbb{R}$ of real numbers results from filling the gaps in the set $\mathbb{Q}$ of rational numbers with *irrational* numbers, such as $\sqrt{2}$. This innovation enables us to consider $\mathbb{R}$ as a *line*, because it has no gaps and the numbers in it are ordered just as we imagine points on a line to be. We say that $\mathbb{R}$, together with its ordering, is a *model* of the line. One of our goals in this chapter is to use $\mathbb{R}$ to build a model for all of Euclidean plane geometry: a structure containing "lines," "circles," "line segments," and so on, with all of the properties required by Euclid's or Hilbert's axioms.

The first step is to build the "plane," and in this we are guided by the properties of parallels in Euclid's geometry. We imagine a pair of perpendicular lines, called the *x-axis* and the *y-axis*, intersecting at a point O called the *origin* (Figure 3.1). We interpret the axes as number lines, with O the number 0 on each, and we assume that the positive direction on the x-axis is to the right and that the positive direction on the y-axis is upward.

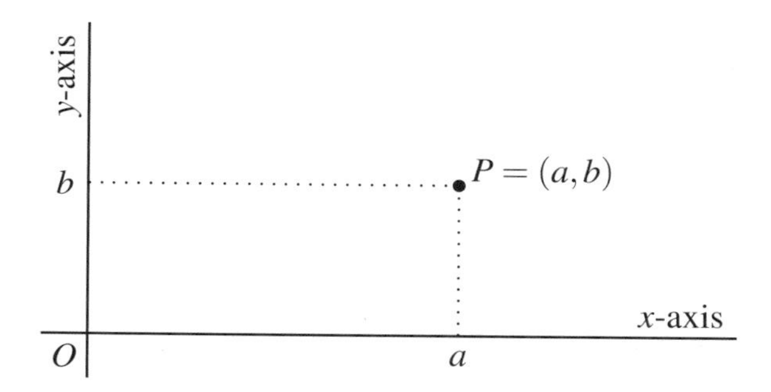

Figure 3.1: Axes and coordinates

Through any point P, there is (by the parallel axiom) a unique line parallel to the y-axis and a unique line parallel to the x-axis. These two lines meet the x-axis and y-axis at numbers a and b called the x- and y-*coordinates* of P, respectively. It is important to remember which number is on the x-axis and which is on the y-axis, because obviously the point with x-coordinate $= 3$ and y-coordinate $= 4$ is different from the point with x-coordinate $= 4$ and y-coordinate $= 3$ (just as the intersection of 3rd Street and 4th Avenue is different from the intersection of 4th Street and 3rd Avenue).

To keep the x-coordinate a and the y-coordinate b in their places, we use the *ordered pair* (a,b). For example, $(3,4)$ is the point with x-coordinate $=3$ and y-coordinate $=4$, whereas $(4,3)$ is the point with x-coordinate $=4$ and y-coordinate $=3$. The ordered pair (a,b) specifies P uniquely because any other point will have at least one different parallel passing through it and hence will differ from P in either the x- or y-coordinate.

Thus, given the existence of a *number line* $\mathbb{R}$ whose points are real numbers, we also have a *number plane* whose points are ordered pairs of real numbers. We often write this number plane as $\mathbb{R} \times \mathbb{R}$ or $\mathbb{R}^2$.

3.2 Lines and their equations

As mentioned in Chapter 2, one of the most important consequences of the parallel axiom is the Thales theorem and hence the proportionality of similar triangles. When coordinates are introduced, this allows us to define the property of straight lines known as *slope*. You know from high-school mathematics that slope is the quotient "rise over run" and, more importantly, that the value of the slope does not depend on which two points of the line define the rise and the run. Figure 3.2 shows why.

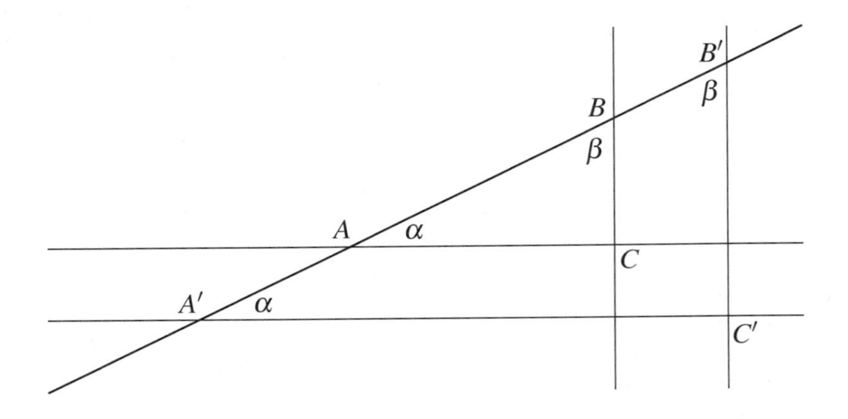

Figure 3.2: Why the slope of a line is constant

In this figure, we have two segments of the same line:

- AB, for which the rise is $|BC|$ and the run is $|AC|$, and
- $A'B'$, for which the rise is $|B'C'|$ and the run is $|A'C'|$.

The angles marked α are equal because AC and $A'C'$ are parallel, and the angles marked β are equal because the BC and $B'C'$ are parallel. Also, the angles at C and C' are both right angles.

Thus, triangles ABC and $A'B'C'$ are similar, and so their corresponding sides are proportional. In particular,

$$\frac{|BC|}{|AC|} = \frac{|B'C'|}{|A'C'|},$$

that is, slope = constant.

Now suppose we are given a line of slope a that crosses the y-axis at the point Q where $y = c$ (Figure 3.3). If $P = (x, y)$ is any point on this line, then the rise from Q to P is $y - c$ and the run is x. Hence

$$\text{slope} = a = \frac{y-c}{x}$$

and therefore, multiplying both sides by x, $y - c = ax$, that is,

$$y = ax + c.$$

This equation is satisfied by all points on the line, and only by them, so we call it the *equation of the line*.

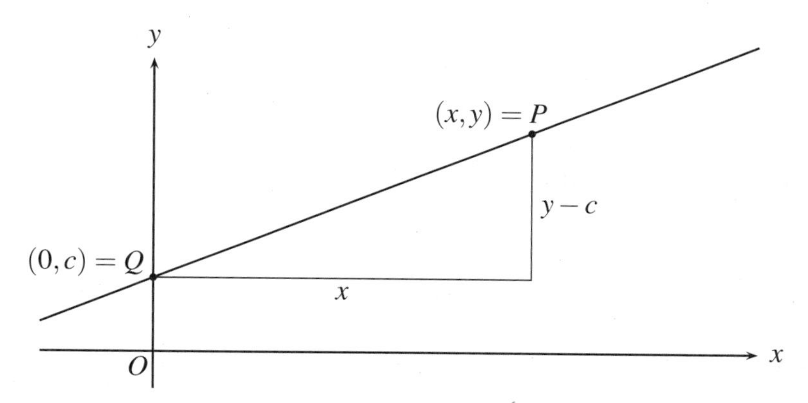

Figure 3.3: Typical point on the line

Almost all lines have equations of this form; the only exceptions are lines that do not cross the y-axis. These are the vertical lines, which also do not have a slope as we have defined it, although we could say they have *infinite slope*. Such a line has an equation of the form

$$x = c, \quad \text{for some constant } c.$$

Thus, all lines have equations of the form

$$ax + by + c = 0, \quad \text{for some constants } a, b, \text{ and } c,$$

called a *linear* equation in the variables x and y.

Up to this point we have been following the steps of Descartes, who viewed equations of lines as *information deduced from Euclid's axioms* (in particular, from the parallel axiom). It is true that Euclid's axioms prompt us to describe lines by linear equations, but we can also take the opposite view: Equations *define* what lines and curves are, and they provide a *model* of Euclid's axioms—showing that geometry follows from properties of the real numbers.

In particular, if a line is defined to be the set of points (x,y) in the number plane satisfying a linear equation then we can prove the following statements that Euclid took as axioms:

- there is a unique line through any two distinct points,
- for any line $\mathscr{L}$ and point P outside $\mathscr{L}$, there is a unique line through P not meeting $\mathscr{L}$.

Because these statements are easy to prove, we leave them to the exercises.

Exercises

Given distinct points $P_1 = (x_1, y_1)$ and $P_2 = (x_2, y_2)$, suppose that $P = (x,y)$ is any point on a line through P_1 and P_2.

3.2.1 By equating slopes, show that x and y satisfy the equation

$$\frac{y_2 - y_1}{x_2 - x_1} = \frac{y - y_1}{x - x_1} \quad \text{if} \quad x_2 \neq x_1.$$

3.2.2 Explain why the equation found in Exercise 3.2.1 is the equation of a straight line.

3.2.3 What happens if $x_2 = x_1$?

Parallel lines, not surprisingly, turn out to be lines with the *same slope*.

3.2.4 Show that distinct lines $y = ax + c$ and $y = a'x + c'$ have a common point unless they have the same slope ($a = a'$). Show that this is also the case when one line has infinite slope.

3.2.5 Deduce from Exercise 3.2.4 that the parallel to a line $\mathscr{L}$ is the unique line through P with the same slope as $\mathscr{L}$.

3.2.6 If $\mathscr{L}$ has equation $y = 3x$, what is the equation of the parallel to $\mathscr{L}$ through $P = (2,2)$?

3.3 Distance

We introduce the concept of *distance* or *length* into the number plane $\mathbb{R}^2$ much as we introduce lines. First we see what Euclid's geometry *suggests* distance should mean; then we turn around and take the suggested meaning as a definition.

Suppose that $P_1 = (x_1, y_1)$ and $P_2 = (x_2, y_2)$ are any two points in $\mathbb{R}^2$. Then it follows from the meaning of coordinates that there is a right-angled triangle as shown in Figure 3.4, and that $|P_1P_2|$ is the length of its hypotenuse.

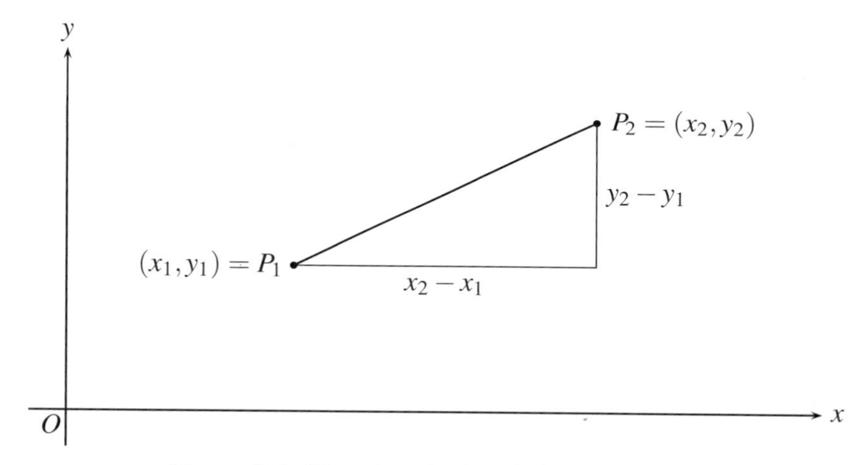

Figure 3.4: The triangle that defines distance

The vertical side of the triangle has length $y_2 - y_1$, and the horizontal side has length $x_2 - x_1$. Then it follows from the Pythagorean theorem that

$$|P_1P_2|^2 = (x_2 - x_1)^2 + (y_2 - y_1)^2,$$

and therefore,

$$|P_1P_2| = \sqrt{(x_2 - x_1)^2 + (y_2 - y_1)^2}. \qquad (*)$$

Thus, it is sensible to *define* the distance $|P_1P_2|$ between any two points P_1 and P_2 by the formula (*). If we do this, the Pythagorean theorem is virtually "true by definition." It is certainly true when the right-angled triangle has a vertical side and a horizontal side, as in Figure 3.4. And we will see later how to rotate any right-angled triangle to such a position (without changing the lengths of its sides).

The equation of a circle

The distance formula (*) leads immediately to the equation of a circle, as follows. Suppose we have a circle with radius r and center at the point $P=(a,b)$. Then any point $Q=(x,y)$ on the circle is at distance r from P, and hence formula (*) gives:

$$r=|PQ|=\sqrt{(x-a)^2+(y-b)^2}.$$

Squaring both sides, we get

$$(x-a)^2+(y-b)^2=r^2.$$

We call this the *equation of the circle* because it is satisfied by any point (x,y) on the circle, and only by such points.

The equidistant line of two points

A circle is the set of points equidistant from a point—its center. It is also natural to ask: What is the set of points equidistant from *two* points in $\mathbb{R}^2$? Answer: *The set of points equidistant from two points is a line.*

To see why, let the two points be $P_1=(a_1,b_1)$ and $P_2=(a_2,b_2)$. Then a point $P=(x,y)$ is equidistant from P_1 and P_2 if $|PP_1|=|PP_2|$, that is, if x and y satisfy the equation

$$\sqrt{(x-a_1)^2+(y-b_1)^2}=\sqrt{(x-a_2)^2+(y-b_2)^2}.$$

Squaring both sides of this equation, we get

$$(x-a_1)^2+(y-b_1)^2=(x-a_2)^2+(y-b_2)^2.$$

Expanding the squares gives

$$x^2-2a_1x+a_1^2+y^2-2b_1y+b_1^2=x^2-2a_2x+a_2^2+y^2-2b_2y+b_2^2.$$

The important thing is that the x^2 and y^2 terms now cancel, which leaves the *linear* equation

$$2(a_2-a_1)x+2(b_2-b_1)y+(b_1^2-b_2^2)=0.$$

Thus, the points $P=(x,y)$ equidistant from P_1 and P_2 form a line. □

Exercises

An interesting application of equidistant lines is the following.

3.3.1 Show that any three points not in a line lie on a unique circle. (Hint: the center of the circle is equidistant from the three points.)

The equations of lines and circles enable us to prove many geometric theorems by algebra, as Descartes realized. In fact, they greatly expand the scope of geometry by allowing many curves to be described by equations. But algebra is also useful in proving that certain quantities are *not* equal. One example is the *triangle inequality*.

3.3.2 Consider any triangle, which for convenience we take to have one vertex at $O = (0,0)$, one at $P = (x_1, 0)$ with $x_1 > 0$, and one at $Q = (x_2, y_2)$. Show that

$$|OP| = x_1, \quad |PQ| = \sqrt{(x_2 - x_1)^2 + y_2^2}, \quad |OQ| = \sqrt{x_2^2 + y_2^2}.$$

The triangle inequality states that $|OP| + |PQ| > |OQ|$ (any two sides of a triangle are together greater than the third). To prove this statement, it suffices to show that

$$(|OP| + |PQ|)^2 > |OQ|^2.$$

3.3.3 Show that $(|OP| + |PQ|)^2 - |OQ|^2 = 2x_1 \left[\sqrt{(x_2 - x_1)^2 + y_2^2} - (x_2 - x_1)\right].$

3.3.4 Show that the term in square brackets in Exercise 3.3.3 is positive if $y_2 \neq 0$, and hence that the triangle inequality holds in this case.

3.3.5 If $y_2 = 0$, why is this not a problem?

Later we will give a more sophisticated approach to the triangle inequality, which does not depend on choosing a special position for the triangle.

3.4 Intersections of lines and circles

Now that lines and circles are defined by equations, we can give exact algebraic equivalents of straightedge and compass operations:

- Drawing a line through given points corresponds to finding the equation of the line through given points (x_1, y_1) and (x_2, y_2). The slope between these two points is $\frac{y_2 - y_1}{x_2 - x_1}$, which must equal the slope $\frac{y - y_1}{x - x_1}$ between the general point (x, y) and the special point (x_1, y_1), so the equation is

$$\frac{y - y_1}{x - x_1} = \frac{y_2 - y_1}{x_2 - x_1}.$$

Multiplying both sides by $(x-x_1)(x_2-x_1)$, we get the equivalent equation

$$(y-y_1)(x_2-x_1) = (x-x_1)(y_2-y_1),$$

or

$$(y_2-y_1)x - (x_2-x_1)y - x_1y_2 + y_1x_2 = 0.$$

- Drawing a circle with given center and radius corresponds to finding the equation of the circle with given center (a,b) and given radius r, which is

$$(x-a)^2 + (y-b)^2 = r^2.$$

- Finding new points as intersections of previously drawn lines and circles corresponds to finding the solution points of
  - a pair of equations of lines,
  - a pair of equations of circles,
  - the equation of a line and the equation of a circle.

For example, to find the intersection of the two circles

$$(x-a_1)^2 + (y-b_1)^2 = r_1^2$$

and

$$(x-a_2)^2 + (y-b_2)^2 = r_2^2,$$

we expand the equations of the circles as

$$x^2 - 2a_1x + a_1^2 + y^2 - 2b_1y + b_1^2 - r_1^2 = 0, \tag{1}$$

$$x^2 - 2a_2x + a_2^2 + y^2 - 2b_2y + b_2^2 - r_2^2 = 0, \tag{2}$$

and subtract Equation (2) from Equation (1). The x^2 and y^2 terms cancel, and we are left with the linear equation in x and y:

$$2(a_2-a_1)x + 2(b_2-b_2)y + r_2^2 - r_1^2 = 0. \tag{3}$$

We can solve Equation (3) for either x or y. Then substituting the solution of (3) in (1) gives a quadratic equation for either y or x. If the equation is of the form $Ax^2 + Bx + C = 0$, then we know that the solutions are

$$x = \frac{-B \pm \sqrt{B^2 - 4AC}}{2A}.$$

Solving linear equations requires only the operations $+, -, \times$, and $\div$, and the quadratic formula shows that $\sqrt{}$ is the only additional operation needed to solve quadratic equations.

Thus, all intersection points involved in a straightedge and compass construction can be found with the operations $+, -, \times, \div$, and $\sqrt{}$.

Now recall from Chapters 1 and 2 that the operations $+, -, \times, \div$, and $\sqrt{}$ can be carried out by straightedge and compass. Hence, we get the following result:

Algebraic criterion for constructibility. *A point is constructible (starting from the points 0 and 1) if and only if its coordinates are obtainable from the number 1 by the operations* $+, -, \times, \div$, *and* $\sqrt{}$.

The algebraic criterion for constructibility was discovered by Descartes, and its greatest virtue is that it enables us to prove that certain figures or points are *not* constructible. For example, one can prove that the number $\sqrt[3]{2}$ is not constructible by showing that it cannot be expressed by a finite number of square roots, and one can prove that the angle $\pi/3$ cannot be trisected by showing that $\cos\frac{\pi}{9}$ also cannot be expressed by a finite number of square roots. These results were not proved until the 19th century, by Pierre Wantzel. Rather sophisticated algebra is required, because one has to go beyond Descartes' concept of constructibility to survey the *totality* of constructible numbers.

Exercises

3.4.1 Find the intersections of the circles $x^2+y^2=1$ and $(x-1)^2+(y-2)^2=4$.

3.4.2 Check the plausibility of your answer to Exercise 3.4.1 by a sketch of the two circles.

3.4.3 The line $x+2y-1=0$ found by eliminating the x^2 and y^2 from the equations of the circles should have some geometric meaning. What is it?

3.5 Angle and slope

The concept of distance is easy to handle in coordinate geometry because the distance between points (x_1,y_1) and (x_2,y_2) is an algebraic function of their coordinates, namely

$$\sqrt{(x_2-x_1)^2+(y_2-y_1)^2}.$$

This is *not* the case for the concept of angle. The angle θ between a line $y = tx$ and the x-axis is $\tan^{-1} t$, and the function $\tan^{-1} t$ is not an algebraic function. Nor is its inverse function $t = \tan\theta$ or the related functions $\sin\theta$ (sine) and $\cos\theta$ (cosine).

To stay within the world of algebra, we have to work with the slope t rather than the angle θ. Lines make the same angle with the x-axis if they have the same slope, but to test equality of angles in general we need the concept of *relative slope*: If line $\mathscr{L}_1$ has slope t_1 and line $\mathscr{L}_2$ has slope t_2, then the *slope of* $\mathscr{L}_1$ *relative to* $\mathscr{L}_2$ is defined to be

$$\pm\left|\frac{t_1 - t_2}{1 + t_1 t_2}\right|.$$

This awkward definition comes from the formula you have probably seen in trigonometry,

$$\tan(\theta_1 - \theta_2) = \frac{\tan\theta_1 - \tan\theta_2}{1 + \tan\theta_1 \tan\theta_2},$$

by taking $t_1 = \tan\theta_1$ and $t_2 = \tan\theta_2$. The reason for the $\pm$ sign and the absolute value is that the slopes t_1, t_2 alone do not specify an angle—they specify only a pair of lines and hence a pair of angles that add to a straight angle. (For more on using relative slope to discuss equality of angles, see Hartshorne's *Geometry: Euclid and Beyond*, particularly pp. 141–155.)

At any rate, with some care it is possible to use the concept of relative slope to test algebraically whether angles are equal. The concept also makes it possible to state the SAS and ASA axioms in coordinate geometry, and to verify that all of Euclid's and Hilbert's axioms hold. We omit the details because they are laborious, and because we can approach SAS and ASA differently now that we have coordinates. Specifically, *it becomes possible to define the concept of "motion" that Euclid appealed to in his proof of SAS!* This will be done in the next section.

Exercises

The most useful instance of relative slope is where the lines are perpendicular.

3.5.1 Show that lines of slopes t_1 and t_2 are perpendicular just in case $t_1 t_2 = -1$.

3.5.2 Use the condition for perpendicularity found in Exercise 3.5.1 to show that the line from $(1,0)$ to $(3,4)$ is perpendicular to the line from $(0,2)$ to $(4,0)$.

In the next section, we will define a *rotation about O* to be a transformation $r_{c,s}$ of $\mathbb{R}^2$ depending on two real numbers c and s such that $c^2+s^2=1$. The transformation $r_{c,s}$ sends the point (x,y) to the point $(cx-sy, sx+cy)$. It will be explained in the next section why it is reasonable to call this a "rotation about O," and why $c=\cos\theta$ and $s=\sin\theta$, where θ is the angle of rotation.

For the moment, suppose that this is the case, and consider the effect of two rotations r_{c_1,s_1} and r_{c_2,s_2}, where

$$c_1=\cos\theta_1,\quad s_1=\sin\theta_1;\qquad c_2=\cos\theta_2,\quad s_2=\sin\theta_2.$$

This thought experiment leads us to proofs of the formulas for cos, sin, and tan of $\theta_1+\theta_2$:

3.5.3 Show that the outcome of r_{c_1,s_1} and r_{c_2,s_2} is to send (x,y) to

$$((c_1c_2-s_1s_2)x-(s_1c_2+c_1s_2)y, (s_1c_2+c_1s_2)x+(c_1c_2-s_1s_2)y).$$

3.5.4 Assuming that r_{c_1,s_1} really is a rotation about O through angle θ_1, and r_{c_2,s_2} really is a rotation about O through angle θ_2, deduce from Exercise 3.5.3 that

$$\begin{aligned}\cos(\theta_1+\theta_2)&=\cos\theta_1\cos\theta_2-\sin\theta_1\sin\theta_2,\\ \sin(\theta_1+\theta_2)&=\sin\theta_1\cos\theta_2+\cos\theta_1\sin\theta_2.\end{aligned}$$

3.5.5 Deduce from Exercise 3.5.4 that

$$\tan(\theta_1+\theta_2)=\frac{\tan\theta_1+\tan\theta_2}{1-\tan\theta_1\tan\theta_2},$$

hence

$$\tan(\theta_1-\theta_2)=\frac{\tan\theta_1-\tan\theta_2}{1+\tan\theta_1\tan\theta_2}.$$

3.6 Isometries

A possible weakness of our model of the plane is that it seems to single out a particular point (the origin O) and particular lines (the x- and y-axes). In Euclid's plane, each point is like any other point and each line is like any other line. We can overcome the apparent bias of $\mathbb{R}^2$ by considering *transformations* that allow any point to become the origin and any line to become the x-axis. As a bonus, this idea gives meaning to the idea of "motion" that Euclid tried to use in his attempt to prove SAS.

A transformation of the plane is simply a function $f:\mathbb{R}^2\rightarrow\mathbb{R}^2$, in other words, a function that sends points to points.

A transformation f is called an *isometry* (from the Greek for "same length") if it sends any two points, P_1 and P_2, to points $f(P_1)$ and $f(P_2)$ the same distance apart. Thus, an isometry is a function f with the property

$$|f(P_1)f(P_2)| = |P_1P_2|$$

for any two points P_1, P_2. Intuitively speaking, an isometry "moves the plane rigidly" because it preserves the distance between points. There are many isometries of the plane, but they can be divided into a few simple and obvious types. We show examples of each type below, and, in the next section, we explain why only these types exist.

You will notice that certain isometries (translations and rotations) make it possible to move the origin to any point in the plane and the x-axis to any line. Thus, $\mathbb{R}^2$ is really like Euclid's plane, in the sense that each point is like any other point and each line is like any other line. This property entitles us to choose axes wherever it is convenient. For example, we are entitled to prove the triangle inequality, as suggested in the Exercises to Section 3.3, by choosing one vertex of the triangle at O and another on the positive x-axis.

Translations

A translation moves each point of the plane the same distance in the same direction. Each translation depends on two constants a and b, so we denote it by $t_{a,b}$. It sends each point (x,y) to the point $(x+a, y+b)$. It is obvious that a translation preserves the distance between any two points, but it is worth checking this formally—so as to know what to do in less obvious cases.

So let $P_1 = (x_1, y_1)$ and $P_2 = (x_2, y_2)$. It follows that

$$t_{a,b}(P_1) = (x_1 + a, y_1 + b), \quad t_{a,b}(P_2) = (x_2 + a, y_2 + b)$$

and therefore,

$$\begin{aligned} |t_{a,b}(P_1)t_{a,b}(P_2)| &= \sqrt{(x_2+a-x_1-a)^2 + (y_2+b-y_1-b)^2} \\ &= \sqrt{(x_2-x_1)^2 + (y_2-y_1)^2} \\ &= |P_1P_2|, \quad \text{as required.} \end{aligned}$$

Rotations

We think of a rotation as something involving an angle θ, but, as mentioned in the previous section, it is more convenient to work algebraically with $\cos\theta$ and $\sin\theta$. These are simply two numbers c and s such that $c^2+s^2=1$, so we will denote a rotation of the plane about the origin by $r_{c,s}$.

The rotation $r_{c,s}$ sends the point (x,y) to the point $(cx-sy, sx+cy)$. It is not obvious why this transformation should be called a rotation, but it becomes clearer after we check that $r_{c,s}$ preserves lengths.

If we let $P_1=(x_1,y_1)$ and $P_2=(x_2,y_2)$ again, it follows that

$$r_{c,s}(P_1)=(cx_1-sy_1, sx_1+cy_1), \quad r_{c,s}(P_2)=(cx_2-sy_2, sx_2+cy_2)$$

and therefore,

$$\begin{aligned}
|r_{c,s}(P_1)r_{c,s}(P_2)| &= \sqrt{[c(x_2-x_1)-s(y_2-y_1)]^2+[s(x_2-x_1)+c(y_2-y_1)]^2} \\
&= \sqrt{\begin{array}{c} c^2(x_2-x_1)^2-2cs(x_2-x_1)(y_2-y_1)+s^2(y_2-y_1)^2 \\ +s^2(x_2-x_1)^2+2cs(x_2-x_1)(y_2-y_1)+c^2(y_2-y_1)^2 \end{array}} \\
&= \sqrt{(c^2+s^2)(x_2-x_1)^2+(c^2+s^2)(y_2-y_1)^2} \\
&= \sqrt{(x_2-x_1)^2+(y_2-y_1)^2} \quad \text{because} \quad c^2+s^2=1 \\
&= |P_1P_2|.
\end{aligned}$$

Thus, $r_{c,s}$ preserves lengths. Also, $r_{c,s}$ sends $(0,0)$ to itself, and it moves $(1,0)$ to (c,s) and $(0,1)$ to $(-s,c)$, *which is exactly what rotation about O through angle θ does* (see Figure 3.5). We will see in the next section that only one isometry of the plane moves these three points in this manner.

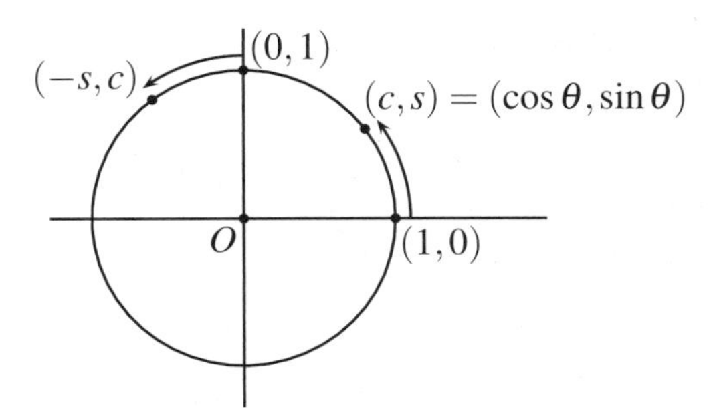

Figure 3.5: Movement of points by a rotation

Reflections

The easiest reflection to describe is *reflection in the x-axis*, which sends $P = (x,y)$ to $\overline{P} = (x,-y)$. Again it is obvious that this is an isometry, but we can check by calculating the distance between reflected points $\overline{P_1}$ and $\overline{P_2}$ (Exercise 3.6.1).

We can reflect the plane in any line, and we can do this by combining reflection in the x-axis with translations and rotations. For example, reflection in the line $y = 1$ (which is parallel to the x-axis) is the result of the following three isometries:

- $t_{0,-1}$, a translation that moves the line $y = 1$ to the x-axis,
- reflection in the x-axis,
- $t_{0,1}$, which moves the x-axis back to the line $y = 1$.

In general, we can do a reflection in any line $\mathscr{L}$ by moving $\mathscr{L}$ to the x-axis by some combination of translation and rotation, reflecting in the x-axis, and then moving the x-axis back to $\mathscr{L}$.

Reflections are the most fundamental isometries, because any isometry is a combination of them, as we will see in the next section. In particular, any translation is a combination of two reflections, and any rotation is a combination of two reflections (see Exercises 3.6.2–3.6.4).

Glide reflections

A glide reflection is the result of a reflection followed by a translation in the direction of the line of reflection. For example, if we reflect in the x-axis, sending (x,y) to $(x,-y)$, and follow this with the translation $t_{1,0}$ of length 1 in the x-direction, then (x,y) ends up at $(x+1,-y)$.

A glide reflection with nonzero translation length is different from the three types of isometry previously considered.

- It is not a translation, because a translation maps any line in the direction of translation into itself, whereas a glide reflection maps only one line into itself (namely, the line of reflection).
- It is not a rotation, because a rotation has a fixed point and a glide reflection does not.
- It is not a reflection, because a reflection also has fixed points (all points on the line of reflection).

Exercises

3.6.1 Check that reflection in the x-axis preserves the distance between any two points.

When we combine reflections in two lines, the nature of the outcome depends on whether the lines are parallel.

3.6.2 Reflect the plane in the x-axis, and then in the line $y = 1/2$. Show that the resulting isometry sends (x, y) to $(x, y+1)$, so it is the translation $t_{0,1}$.

3.6.3 Generalize the idea of Exercise 3.6.2 to show that the combination of reflections in parallel lines, distance $d/2$ apart, is a translation through distance d, in the direction perpendicular to the lines of reflection.

3.6.4 Show, by a suitable picture, that the combination of reflections in lines that meet at angle $\theta/2$ is a rotation through angle θ, about the point of intersection of the lines.

Another way to put the result of Exercise 3.6.4 is as follows: Reflections in *any* two lines meeting at the same angle $\theta/2$ at the same point P give the same outcome. This observation is important for the next three exercises (where pictures will also be helpful).

3.6.5 Show that reflections in lines $\mathscr{L}$, $\mathscr{M}$, and $\mathscr{N}$ (in that order) have the same outcome as reflections in lines $\mathscr{L}'$, $\mathscr{M}'$, and $\mathscr{N}$, where $\mathscr{M}'$ is perpendicular to $\mathscr{N}$.

3.6.6 Next show that reflections in lines $\mathscr{L}'$, $\mathscr{M}'$, and $\mathscr{N}$ have the same outcome as reflections in lines $\mathscr{L}'$, $\mathscr{M}''$, and $\mathscr{N}'$, where $\mathscr{M}''$ is parallel to $\mathscr{L}'$ and $\mathscr{N}'$ is perpendicular to $\mathscr{M}''$.

3.6.7 Deduce from Exercise 3.6.6 that the combination of any three reflections is a glide reflection.

3.7 The three reflections theorem

We saw in Section 3.3 that the points equidistant from two points A and B form a line, which implies that isometries of the plane are very simple: *An isometry f of $\mathbb{R}^2$ is determined by the images $f(A), f(B), f(C)$ of three points A, B, C not in a line.*

The proof follows from three simple observations:

- Any point P in $\mathbb{R}^2$ is determined by its distances from A, B, C. Because if Q is another point with the same distances from A, B, C as P, then A, B, C lie in the equidistant line of P and Q, contrary to the assumption that A, B, C are *not* in a line.

- The isometry f preserves distances (by definition of isometry), so $f(P)$ lies at the same respective distances from $f(A), f(B), f(C)$ as P does from A, B, C.
- There is only one point at given distances from $f(A), f(B), f(C)$ because these three points are not in a line—in fact they form a triangle congruent to triangle ABC, because f preserves distances.

Thus, the image $f(P)$ of any point P—and hence the whole isometry f—is determined by the images of three points A, B, C not in a line. □

This "three point determination theorem" gives us the:

Three reflections theorem. *Any isometry of $\mathbb{R}^2$ is a combination of one, two, or three reflections.*

Given an isometry f, we choose three points A, B, C not in a line, and we look for a combination of reflections that sends A to $f(A)$, B to $f(B)$, and C to $f(C)$. Such a combination is necessarily equal to f. We can certainly send A to $f(A)$ by reflection in the equidistant line of A and $f(A)$. Call this reflection r_A.

Now r_A sends B to $r_A(B)$, so if $r_A(B) = f(B)$ we need to do nothing more for B.

If $r_A(B) \neq f(B)$, we can send $r_A(B)$ to $f(B)$ by reflection r_B in the equidistant line of $r_A(B)$ and $f(B)$. Fortunately, $f(A) = r_A(A)$ lies on this line, because the distance from $f(A)$ to $f(B)$ equals the distance from $r_A(A)$ to $r_A(B)$ (because f and r_A are isometries). Thus, r_B does not move $f(A)$, and the *combination* of r_A followed by r_B sends A to $f(A)$ and B to $f(B)$.

The argument is similar for C. If C has already been sent to $f(C)$, we are done. If not, we reflect in the line equidistant from $f(C)$ and the point where C has been sent so far. It turns out (by a check of equal distances like that made for $f(A)$ above) that $f(A)$ and $f(B)$ already lie on this line, so they are not moved. Thus, we finally have a combination of no more than three reflections that moves A to $f(A)$, B to $f(B)$, and C to $f(C)$, as required. □

Now of course, one reflection is a reflection, and we found in the previous exercise set that combinations of two reflections are translations and rotations, and that combinations of three reflections are glide reflections (which include reflections). Thus, *an isometry of $\mathbb{R}^2$ is either a translation, a rotation, or a glide reflection.*

Exercises

Given three points A, B, C and the points $f(A), f(B), f(C)$ to which they are sent by an isometry f, it is possible to find three reflections that combine to form f by following the steps in the proof above. However, if one merely wants to know what *kind* of isometry f is—translation, rotation, or glide reflection—then the answer can be found more simply.

To fix ideas, we take the initial three points to be $A = (0,1)$, $B = (0,0)$, and $C = (1,0)$. You will probably find it helpful to sketch the triples of points $f(A)$, $f(B)$, $f(C)$ given in the following exercises.

3.7.1 Suppose that $f(A) = (1.4,2)$, $f(B) = (1.4,1)$, and $f(C) = (2.4,1)$. Is f a translation or a rotation? How can you tell that f is not a glide reflection?

3.7.2 Suppose that $f(A) = (0.4,1.8)$, $f(B) = (1,1)$, and $f(C) = (1.8,1.6)$. We can tell that f is not a translation or glide reflection (hence, it must be a rotation). How?

3.7.3 Suppose that $f(A) = (1.8,1.6)$, $f(B) = (1,1)$, and $f(C) = (0.4,1.8)$. How do I know that this is a glide reflection?

3.7.4 State a simple test for telling whether f is a translation, rotation, or glide reflection from the positions of $f(A)$, $f(B)$, and $f(C)$.

3.8 Discussion

The discovery of coordinates is rightly considered a turning point in the development of mathematics because it reveals a vast new panorama of geometry, open to exploration in at least three different directions.

- Description of curves by equations, and their analysis by algebra. This direction is called *algebraic geometry*, and the curves described by polynomial equations are called *algebraic curves*. Straight lines, described by the linear equations $ax + by + c = 0$, are called curves of *degree 1*. Circles, described by the equations $(x-a)^2 + (y-b)^2 = r^2$, are curves of *degree 2*, and so on.

  One can see that there are curves of arbitrarily high degree, so most of algebraic geometry is beyond the scope of this book. Even the curves of degree 3 are worth a book of their own, so for them, and other algebraic curves, we refer readers elsewhere. Two excellent books, which show how algebraic geometry relates to other parts of mathematics, are *Elliptic Curves* by H. P. McKean and V. Moll and *Plane Algebraic Curves* by E. Brieskorn and H. Knörrer.

- Algebraic study of objects described by linear equations (such as lines and planes). Even this is a big subject, called *linear algebra.* Although it is technically part of algebraic geometry, it has a special flavor, very close to that of Euclidean geometry. We explore plane geometry from the viewpoint of linear algebra in Chapter 4, and later we make some brief excursions into three and four dimensions.

  The real strength of linear algebra is its ability to describe spaces of any number of dimensions in geometric language. Again, this investigation is beyond our scope, but we will recommend additional reading at the appropriate places.

- The study of transformations, which draws on the special branch of algebra known as *group theory.* Because many geometric transformations are described by linear equations, this study overlaps with linear algebra. The role of transformations was first emphasized by the German mathematician Felix Klein, in an address he delivered at the University of Erlangen in 1872. His address, known by its German name the *Erlanger Programm*, characterizes geometry as the study of *transformation groups* and their *invariants*.

So far, we have seen only one transformation group and a handful of invariants—the group of isometries of $\mathbb{R}^2$ and what it leaves invariant (length, angle, straightness)—so the importance of Klein's idea can hardly be clear yet. However, in Chapter 4 we introduce a very different group of transformations and a very different invariant—the *projective transformations* and the *cross-ratio*—so readers are asked to bear with us. In Chapters 7 and 8, we develop Klein's idea in some generality and give another significant example, the geometry of the "non-Euclidean" plane.

4

Vectors and Euclidean spaces

PREVIEW

In this chapter, we process coordinates by *linear algebra*. We view points as *vectors* that can be added and multiplied by numbers, and we introduce the *inner product* of vectors, which gives an efficient algebraic method to deal with both lengths and angles.

We revisit some theorems of Euclid to see where they fit in the world of vector geometry, and we become acquainted with some theorems that are particularly natural in this environment.

For plane geometry, the appropriate vectors are ordered pairs (x, y) of real numbers. We *add* pairs according to the rule

$$(u_1, u_2) + (v_1, v_2) = (u_1 + v_1, u_2 + v_2),$$

and *multiply a pair by a real number a* according to the rule

$$a(u_1, u_2) = (au_1, au_2).$$

These vector operations do not involve the concept of length or distance; yet they enable us to discuss certain ratios of lengths and to prove the theorems of Thales and Pappus.

The concept of distance is introduced through the concept of *inner product* $\mathbf{u} \cdot \mathbf{v}$ of vectors $\mathbf{u}$ and $\mathbf{v}$. If $\mathbf{u} = (u_1, u_2)$ and $\mathbf{v} = (v_1, v_2)$, then

$$\mathbf{u} \cdot \mathbf{v} = u_1 v_1 + u_2 v_2.$$

The inner product gives us distance because $\mathbf{u} \cdot \mathbf{u} = |\mathbf{u}|^2$, where $|\mathbf{u}|$ is the distance of $\mathbf{u}$ from the origin $\mathbf{0}$. It also gives us angle because

$$\mathbf{u} \cdot \mathbf{v} = |\mathbf{u}||\mathbf{v}| \cos \theta,$$

where θ is the angle between the directions of $\mathbf{u}$ and $\mathbf{v}$ from $\mathbf{0}$.

4.1 Vectors

Vectors are mathematical objects that can be added, and multiplied by numbers, subject to certain rules. The real numbers are the simplest example of vectors, and the rules for sums and multiples of any vectors are just the following properties of sums and multiples of numbers:

$$\begin{aligned} \mathbf{u}+\mathbf{v} &= \mathbf{v}+\mathbf{u} & 1\mathbf{u} &= \mathbf{u} \\ \mathbf{u}+(\mathbf{v}+\mathbf{w}) &= (\mathbf{u}+\mathbf{v})+\mathbf{w} & a(\mathbf{u}+\mathbf{v}) &= a\mathbf{u}+a\mathbf{v} \\ \mathbf{u}+\mathbf{0} &= \mathbf{u} & (a+b)\mathbf{u} &= a\mathbf{u}+b\mathbf{u} \\ \mathbf{u}+(-\mathbf{u}) &= \mathbf{0} & a(b\mathbf{u}) &= (ab)\mathbf{u}. \end{aligned}$$

These rules obviously hold when $a, b, 1, \mathbf{u}, \mathbf{v}, \mathbf{w}, \mathbf{0}$ are all numbers, and $\mathbf{0}$ is the ordinary zero.

They also hold when $\mathbf{u}, \mathbf{v}, \mathbf{w}$ are *points in the plane* $\mathbb{R}^2$, if we interpret $\mathbf{0}$ as $(0,0)$, $+$ as the *vector sum* defined for $\mathbf{u} = (u_1, u_2)$ and $\mathbf{v} = (v_1, v_2)$ by

$$(u_1, u_2) + (v_1 + v_2) = (u_1 + v_1, u_2 + v_2),$$

and $a\mathbf{u}$ as the *scalar multiple* defined by

$$a(u_1, u_2) = (au_1, au_2).$$

The vector sum is geometrically interesting, because $\mathbf{u}+\mathbf{v}$ is the fourth vertex of a parallelogram formed by the points $\mathbf{0}$, $\mathbf{u}$, and $\mathbf{v}$ (Figure 4.1).

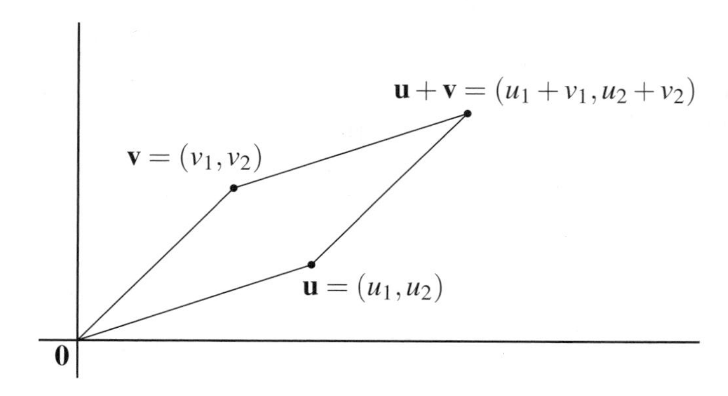

Figure 4.1: The parallelogram rule for vector sum

In fact, the rule for forming the sum of two vectors is often called the "parallelogram rule."

Scalar multiplication by a is also geometrically interesting, because it represents magnification by the factor a. It magnifies, or *dilates*, the whole plane by the factor a, transforming each figure into a similar copy of itself. Figure 4.2 shows an example of this with $a = 2.5$.

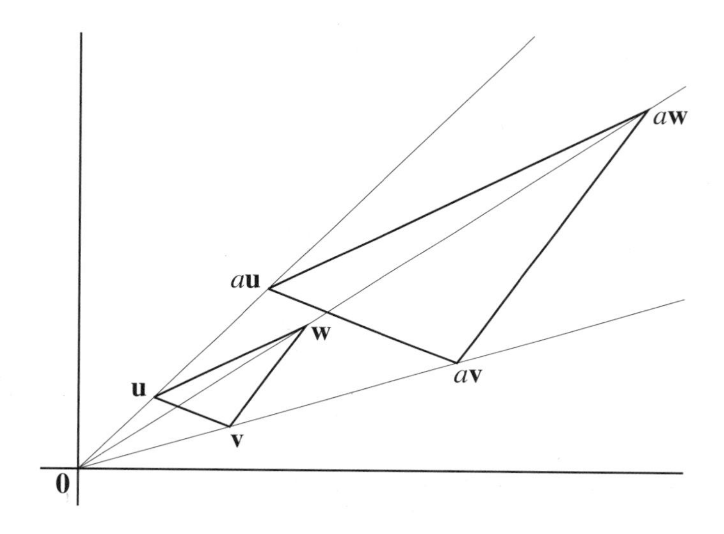

Figure 4.2: Scalar multiplication as a dilation of the plane

Real vector spaces

It seems that the operations of vector addition and scalar multiplication capture some geometrically interesting features of a space. With this in mind, we define a *real vector space* to be a set V of objects, called *vectors*, with operations of vector addition and scalar multiplication satisfying the following conditions:

- If $\mathbf{u}$ and $\mathbf{v}$ are in V, then so are $\mathbf{u}+\mathbf{v}$ and $a\mathbf{u}$ for any real number a.
- There is a *zero vector* $\mathbf{0}$ such that $\mathbf{u}+\mathbf{0}=\mathbf{u}$ for each vector $\mathbf{u}$. Each $\mathbf{u}$ in V has a *additive inverse* $-\mathbf{u}$ such that $\mathbf{u}+(-\mathbf{u})=\mathbf{0}$.
- Vector addition and scalar multiplication on V have the eight properties listed at the beginning of this section.

It turns out that real vector spaces are a natural setting for Euclidean geometry. We must introduce extra structure, which is called the *inner product*, before we can talk about length and angle. But once the inner product is there, we can prove all theorems of Euclidean geometry, often more efficiently than before. Also, we can uniformly extend geometry to *any number of dimensions* by considering the space $\mathbb{R}^n$ of *ordered n-tuples* of real numbers $(x_1, x_2, \ldots, x_n)$.

For example, we can study three-dimensional Euclidean geometry in the space of ordered triples

$$\mathbb{R}^3 = \{(x_1, x_2, x_3) : x_1, x_2, x_3 \in \mathbb{R}\},$$

where the sum of $\mathbf{u} = (u_1, u_2, u_3)$ and $\mathbf{v} = (v_1, v_2, v_3)$ is defined by

$$(u_1, u_2, u_3) + (v_1, v_2, v_3) = (u_1 + v_1, u_2 + v_2, u_3 + v_3)$$

and the scalar multiple $a\mathbf{u}$ is defined by

$$a(u_1, u_2, u_3) = (au_1, au_2, au_3).$$

Exercises

It is obvious that $\mathbb{R}^2$ has the eight properties of a real vector space. However, it is worth noting that $\mathbb{R}^2$ "inherits" these eight properties from the corresponding properties of real numbers. For example, the property $\mathbf{u} + \mathbf{v} = \mathbf{v} + \mathbf{u}$ (called the *commutative law*) for vector addition is inherited from the corresponding commutative law for number addition, $u + v = v + u$, as follows:

$$\begin{aligned}
\mathbf{u} + \mathbf{v} &= (u_1, u_2) + (v_1 + v_2) \\
&= (u_1 + v_1, u_2 + v_2) \quad \text{by definition of vector addition} \\
&= (v_1 + u_1, v_2 + u_2) \quad \text{by commutative law for numbers} \\
&= (v_1, v_2) + (u_1, u_2) \quad \text{by definition of vector addition} \\
&= \mathbf{v} + \mathbf{u}.
\end{aligned}$$

4.1.1 Check that the other seven properties of a vector space for $\mathbb{R}^2$ are inherited from corresponding properties of $\mathbb{R}$.

4.1.2 Similarly check that $\mathbb{R}^n$ has the eight properties of a vector space.

The term "dilation" for multiplication of all vectors in $\mathbb{R}^2$ (or $\mathbb{R}^n$ for that matter) by a real number a goes a little beyond the everyday meaning of the word in the case when a is smaller than 1 or negative.

4.1.3 What is the geometric meaning of the transformation of $\mathbb{R}^2$ when every vector is multiplied by -1? Is it a rotation?

4.1.4 Is it a rotation of $\mathbb{R}^3$ when every vector is multiplied by -1?

4.2 Direction and linear independence

Vectors give a concept of *direction* in $\mathbb{R}^2$ by representing lines through $\mathbf{0}$. If $\mathbf{u}$ is a nonzero vector, then the real multiples $a\mathbf{u}$ of $\mathbf{u}$ make up the line through $\mathbf{0}$ and $\mathbf{u}$, so we call them the points "in direction $\mathbf{u}$ from $\mathbf{0}$." (You may prefer to say that $-\mathbf{u}$ is in the direction *opposite* to $\mathbf{u}$, but it is simpler to associate direction with a whole line, rather than a half line.)

Nonzero vectors $\mathbf{u}$ and $\mathbf{v}$, therefore, have *different directions from* $\mathbf{0}$ if neither is a multiple of the other. It follows that such $\mathbf{u}$ and $\mathbf{v}$ are *linearly independent*; that is, there are no real numbers a and b, not both zero, with

$$a\mathbf{u} + b\mathbf{v} = \mathbf{0}.$$

Because, if one of a, b is not zero in this equation, we can divide by it and hence express one of $\mathbf{u}$, $\mathbf{v}$ as a multiple of the other.

The concept of direction has an obvious generalization: $\mathbf{w}$ *has direction* $\mathbf{u}$ *from* $\mathbf{v}$ (or *relative to* $\mathbf{v}$) if $\mathbf{w} - \mathbf{v}$ is a multiple of $\mathbf{u}$. We also say that "$\mathbf{w} - \mathbf{v}$ has direction $\mathbf{u}$," and there is no harm in viewing $\mathbf{w} - \mathbf{v}$ as an abbreviation for the line segment from $\mathbf{v}$ to $\mathbf{w}$. As in coordinate geometry, we say that line segments from $\mathbf{v}$ to $\mathbf{w}$ and from $\mathbf{s}$ to $\mathbf{t}$ are *parallel* if they have the same direction; that is, if

$$\mathbf{w} - \mathbf{v} = a(\mathbf{t} - \mathbf{s}) \quad \text{for some real number } a \neq 0.$$

Figure 4.3 shows an example of parallel line segments, from $\mathbf{v}$ to $\mathbf{w}$ and from $\mathbf{s}$ to $\mathbf{t}$, both of which have direction $\mathbf{u}$.

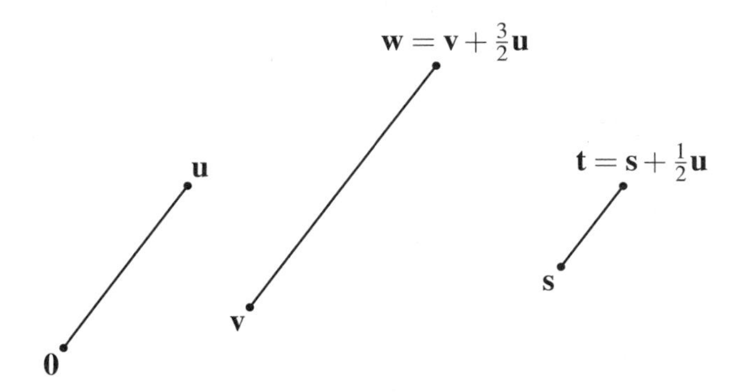

Figure 4.3: Parallel line segments with direction $\mathbf{u}$

Here we have

$$\mathbf{w} - \mathbf{v} = \frac{3}{2}\mathbf{u} \quad \text{and} \quad \mathbf{t} - \mathbf{s} = \frac{1}{2}\mathbf{u}, \quad \text{so} \quad \mathbf{w} - \mathbf{v} = 3(\mathbf{t} - \mathbf{s}).$$

Now let us try out the vector concept of parallels on two important theorems from previous chapters. The first is a version of the Thales theorem that parallels cut a pair of lines in proportional segments.

Vector Thales theorem. *If* $\mathbf{s}$ *and* $\mathbf{v}$ *are on one line through* $\mathbf{0}$, $\mathbf{t}$ *and* $\mathbf{w}$ *are on another, and* $\mathbf{w}-\mathbf{v}$ *is parallel to* $\mathbf{t}-\mathbf{s}$, *then* $\mathbf{v}=a\mathbf{s}$ *and* $\mathbf{w}=a\mathbf{t}$ *for some number* a.

If $\mathbf{w}-\mathbf{v}$ is parallel to $\mathbf{t}-\mathbf{s}$, then

$$\mathbf{w}-\mathbf{v}=a(\mathbf{t}-\mathbf{s})=a\mathbf{t}-a\mathbf{s} \quad \text{for some real number } a.$$

Because $\mathbf{v}$ is on the same line through $\mathbf{0}$ as $\mathbf{s}$, we have $\mathbf{v}=b\mathbf{s}$ for some b, and similarly $\mathbf{w}=c\mathbf{t}$ for some c (this is a good moment to draw a picture). It follows that

$$\mathbf{w}-\mathbf{v}=c\mathbf{t}-b\mathbf{s}=a\mathbf{t}-a\mathbf{s},$$

and therefore,

$$(c-a)\mathbf{t}+(a-b)\mathbf{s}=\mathbf{0}.$$

But $\mathbf{s}$ and $\mathbf{t}$ are in different directions from $\mathbf{0}$, hence linearly independent, so

$$c-a=a-b=0.$$

Thus, $\mathbf{v}=a\mathbf{s}$ and $\mathbf{w}=a\mathbf{t}$, as required. □

As in axiomatic geometry (Exercise 1.4.3), the Pappus theorem follows from the Thales theorem. However, "proportionality" is easier to handle with vectors.

Vector Pappus theorem. *If* $\mathbf{r}$, $\mathbf{s}$, $\mathbf{t}$, $\mathbf{u}$, $\mathbf{v}$, $\mathbf{w}$ *lie alternately on two lines through* $\mathbf{0}$, *with* $\mathbf{u}-\mathbf{v}$ *parallel to* $\mathbf{s}-\mathbf{r}$ *and* $\mathbf{t}-\mathbf{s}$ *parallel to* $\mathbf{v}-\mathbf{w}$, *then* $\mathbf{u}-\mathbf{t}$ *is parallel to* $\mathbf{w}-\mathbf{r}$.

Figure 4.4 shows the situation described in the theorem.

Because $\mathbf{u}-\mathbf{v}$ is parallel to $\mathbf{s}-\mathbf{r}$, we have $\mathbf{u}=a\mathbf{s}$ and $\mathbf{v}=a\mathbf{r}$ for some number a. Because $\mathbf{t}-\mathbf{s}$ is parallel to $\mathbf{v}-\mathbf{w}$, we have $\mathbf{s}=b\mathbf{w}$ and $\mathbf{t}=b\mathbf{v}$ for some number b.

From these two facts, we conclude that

$$\mathbf{u}=a\mathbf{s}=ab\mathbf{w} \quad \text{and} \quad \mathbf{t}=b\mathbf{v}=ba\mathbf{r},$$

hence,

$$\mathbf{u}-\mathbf{t}=ab\mathbf{w}-ba\mathbf{r}=ab(\mathbf{w}-\mathbf{r}),$$

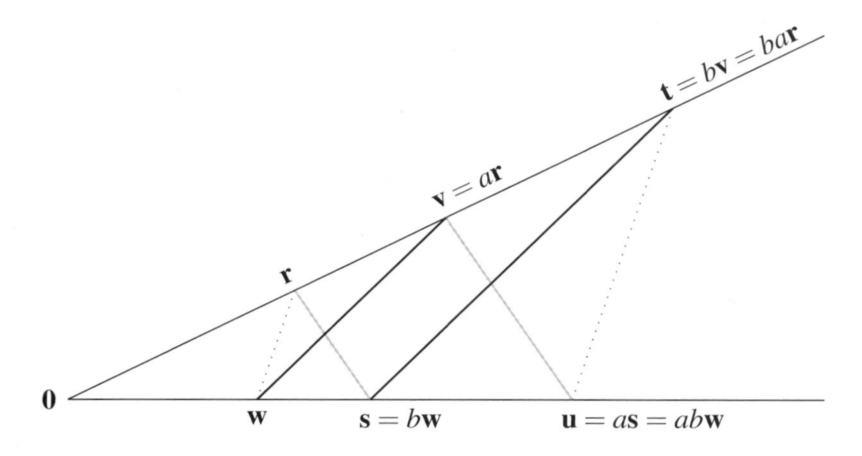

Figure 4.4: The parallel Pappus configuration, labeled by vectors

and therefore, $\mathbf{u}-\mathbf{t}$ is parallel to $\mathbf{w}-\mathbf{r}$. $\square$

The last step in this proof, where we exchange ba for ab, is of course a trifle, because $ab = ba$ for any real numbers a and b. But it is a big step in Chapter 6, where we try to develop geometry without numbers. There we have to build an arithmetic of line segments, and the Pappus theorem is crucial in getting multiplication to behave properly.

Exercises

In Chapter 1, we mentioned that a second theorem about parallels, the Desargues theorem, often appears alongside the Pappus theorem in the foundations of geometry. This situation certainly holds in vector geometry, where the appropriate Desargues theorem likewise follows from the vector Thales theorem.

4.2.1 Following the setup explained in Exercise 1.4.4, and the formulation of the vector Pappus theorem above, formulate a "vector Desargues theorem."

4.2.2 Prove your vector Desargues theorem with the help of the vector Thales theorem.

4.3 Midpoints and centroids

The definition of a real vector space does not include a definition of distance, but we can speak of the midpoint of the line segment from $\mathbf{u}$ to $\mathbf{v}$ and, more generally, of the point that divides this segment in a given ratio.

To see why, first observe that $\mathbf{v}$ is obtained from $\mathbf{u}$ by adding $\mathbf{v}-\mathbf{u}$, the vector that represents the position of $\mathbf{v}$ *relative* to $\mathbf{u}$. More generally, adding any scalar multiple $a(\mathbf{v}-\mathbf{u})$ to $\mathbf{u}$ produces a point whose *direction* relative to $\mathbf{u}$ is the same as that of $\mathbf{v}$. Thus, the points $\mathbf{u}+a(\mathbf{v}-\mathbf{u})$ are precisely those on the line through $\mathbf{u}$ and $\mathbf{v}$. In particular, the midpoint of the segment between $\mathbf{u}$ and $\mathbf{v}$ is obtained by adding $\frac{1}{2}(\mathbf{v}-\mathbf{u})$ to $\mathbf{u}$, and hence,

$$\text{midpoint of line segment between } \mathbf{u} \text{ and } \mathbf{v} = \mathbf{u}+\frac{1}{2}(\mathbf{v}-\mathbf{u}) = \frac{1}{2}(\mathbf{u}+\mathbf{v}).$$

One might describe this result by saying that the midpoint of the line segment between $\mathbf{u}$ and $\mathbf{v}$ is the *vector average* of $\mathbf{u}$ and $\mathbf{v}$.

This description of the midpoint gives a very short proof of the theorem from Exercise 2.2.1, that the diagonals of a parallelogram bisect each other. By choosing one of the vertices of the parallelogram at $\mathbf{0}$, we can assume that the other vertices are at $\mathbf{u}$, $\mathbf{v}$, and $\mathbf{u}+\mathbf{v}$ (Figure 4.5).

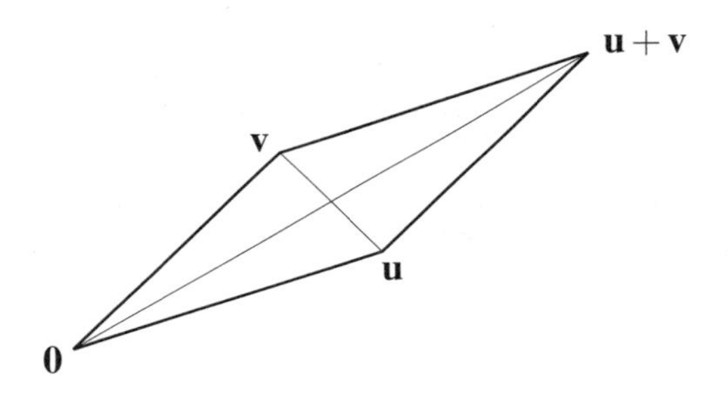

Figure 4.5: Diagonals of a parallelogram

Then the midpoint of the diagonal from $\mathbf{0}$ to $\mathbf{u}+\mathbf{v}$ is $\frac{1}{2}(\mathbf{u}+\mathbf{v})$. And, by the result just proved, this is also the midpoint of the other diagonal—the line segment between $\mathbf{u}$ and $\mathbf{v}$. □

The vector average of two or more points is physically significant because it is the *barycenter* or *center of mass* of the system obtained by placing equal masses at the given points. The geometric name for this vector average point is the *centroid*.

In the case of a triangle, the centroid has an alternative geometric description, given by the following classical theorem about *medians*: the lines from the vertices of a triangle to the midpoints of the respective opposite sides.

Concurrence of medians. *The medians of any triangle pass through the same point, the centroid of the triangle.*

To prove this theorem, suppose that the vertices of the triangle are $\mathbf{u}$, $\mathbf{v}$, and $\mathbf{w}$. Then the median from $\mathbf{u}$ goes to the midpoint $\frac{1}{2}(\mathbf{v}+\mathbf{w})$, and so on, as shown in Figure 4.6.

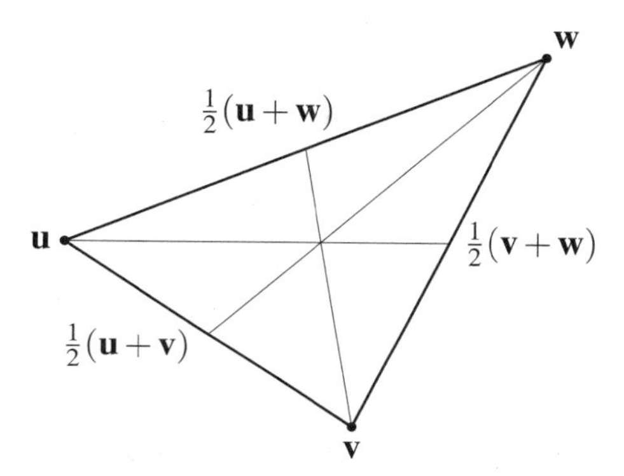

Figure 4.6: The medians of a triangle

Looking at this figure, it seems likely that the medians meet at the point 2/3 of the way from $\mathbf{u}$ to $\frac{1}{2}(\mathbf{v}+\mathbf{w})$, that is, at the point

$$\mathbf{u}+\frac{2}{3}\left(\frac{1}{2}(\mathbf{v}+\mathbf{w})-\mathbf{u}\right)=\mathbf{u}+\frac{1}{3}(\mathbf{v}+\mathbf{w})-\frac{2}{3}\mathbf{u}=\frac{1}{3}(\mathbf{u}+\mathbf{v}+\mathbf{w}).$$

Voilà! This is the centroid, and a similar argument shows that it lies 2/3 of the way between $\mathbf{v}$ and $\frac{1}{2}(\mathbf{u}+\mathbf{w})$ and 2/3 of the way between $\mathbf{w}$ and $\frac{1}{2}(\mathbf{u}+\mathbf{v})$. That is, the centroid is the common point of all three medians.□

You can of course check by calculation that $\frac{1}{3}(\mathbf{u}+\mathbf{v}+\mathbf{w})$ lies 2/3 of the way between $\mathbf{v}$ and $\frac{1}{2}(\mathbf{u}+\mathbf{w})$ and also 2/3 of the way between $\mathbf{w}$ and $\frac{1}{2}(\mathbf{u}+\mathbf{v})$. But the smart thing is not to *do* the calculation but to *predict the result*. We know that calculating the point 2/3 of the way between $\mathbf{u}$ and $\frac{1}{2}(\mathbf{v}+\mathbf{w})$ gives

$$\frac{1}{3}(\mathbf{u}+\mathbf{v}+\mathbf{w}),$$

a result that is unchanged when we permute the letters $\mathbf{u}$, $\mathbf{v}$, and $\mathbf{w}$. The other two calculations are the same, except for the ordering of the letters $\mathbf{u}$, $\mathbf{v}$, and $\mathbf{w}$. Hence, they lead to the same result.

Exercises

4.3.1 Show that a square with vertices **t**, **u**, **v**, **w** has center $\frac{1}{4}(\mathbf{t}+\mathbf{u}+\mathbf{v}+\mathbf{w})$.

The theorem about concurrence of medians generalizes beautifully to three dimensions, where the figure corresponding to a triangle is a *tetrahedron*: a solid with four vertices joined by six lines that bound the tetrahedron's four triangular faces (Figure 4.7).

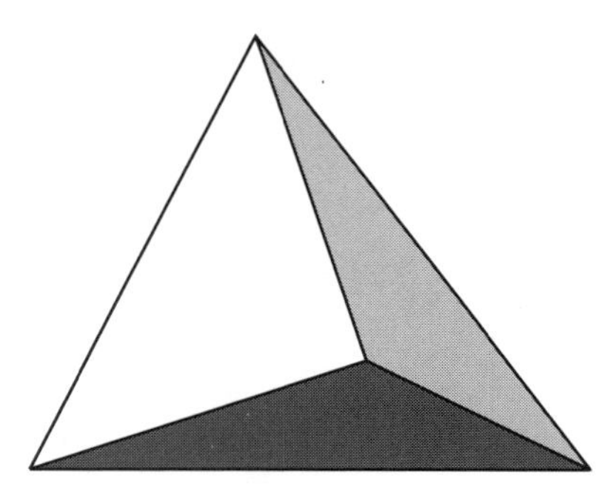

Figure 4.7: A tetrahedron

4.3.2 Suppose that the tetrahedron has vertices **t**, **u**, **v**, and **w**. Show that the centroid of the face opposite to **t** is $\frac{1}{3}(\mathbf{u}+\mathbf{v}+\mathbf{w})$, and write down the centroids of the other three faces.

4.3.3 Now consider each line joining a vertex to the centroid of the opposite face. In particular, show that the point 3/4 of the way from **t** to the centroid of the opposite face is $\frac{1}{4}(\mathbf{t}+\mathbf{u}+\mathbf{v}+\mathbf{w})$—the centroid of the tetrahedron.

4.3.4 Explain why the point $\frac{1}{4}(\mathbf{t}+\mathbf{u}+\mathbf{v}+\mathbf{w})$ lies on the other three lines from a vertex to the centroid of the opposite face.

4.3.5 Deduce that the four lines from vertex to centroid of opposite face meet at the centroid of the tetrahedron.

4.4 The inner product

If $\mathbf{u} = (u_1, u_2)$ and $\mathbf{v} = (v_1, v_2)$ are vectors in $\mathbb{R}^2$, we define their *inner product* $\mathbf{u} \cdot \mathbf{v}$ to be $u_1v_1 + u_2v_2$. Thus, the inner product of two vectors is not another vector, but a real number or "scalar." For this reason, $\mathbf{u} \cdot \mathbf{v}$ is also called the *scalar product* of **u** and **v**.

It is easy to check, from the definition, that the inner product has the algebraic properties

$$\begin{aligned}\mathbf{u}\cdot\mathbf{v}&=\mathbf{v}\cdot\mathbf{u},\\ \mathbf{u}\cdot(\mathbf{v}+\mathbf{w})&=\mathbf{u}\cdot\mathbf{v}+\mathbf{u}\cdot\mathbf{w},\\ (a\mathbf{u})\cdot\mathbf{v}&=\mathbf{u}\cdot(a\mathbf{v})=a(\mathbf{u}\cdot\mathbf{v}),\end{aligned}$$

which immediately give information about length and angle:

- The length $|\mathbf{u}|$ is the distance of $\mathbf{u}=(u_1,u_2)$ from $\mathbf{0}$, which is $\sqrt{u_1^2+u_2^2}$ by the definition of distance in $\mathbb{R}^2$ (Section 3.3). Hence,

$$|\mathbf{u}|^2=u_1^2+u_2^2=\mathbf{u}\cdot\mathbf{u}.$$

It follows that the square of the distance $|\mathbf{v}-\mathbf{u}|$ from $\mathbf{u}$ to $\mathbf{v}$ is

$$|\mathbf{v}-\mathbf{u}|^2=(\mathbf{v}-\mathbf{u})\cdot(\mathbf{v}-\mathbf{u})=|\mathbf{u}|^2+|\mathbf{v}|^2-2\mathbf{u}\cdot\mathbf{v}.$$

- Vectors $\mathbf{u}$ and $\mathbf{v}$ are perpendicular if and only if $\mathbf{u}\cdot\mathbf{v}=0$. Because $\mathbf{u}$ has slope u_2/u_1 and $\mathbf{v}$ has slope v_2/v_1, and we know from Section 3.5 that they are perpendicular if and only the product of their slopes is -1. That means

$$\frac{u_2}{u_1}=-\frac{v_1}{v_2}\quad\text{and hence}\quad u_2v_2=-u_1v_1,$$

multiplying both sides by u_1v_2. This equation holds if and only if

$$0=u_1v_1+u_2v_2=\mathbf{u}\cdot\mathbf{v}.$$

We will see in the next section how to extract more information about angle from the inner product. The formula above for $|\mathbf{v}-\mathbf{u}|^2$ turns out to be the "cosine rule" or "law of cosines" from high-school trigonometry. But even the criterion for perpendicularity gives a simple proof of a far-from-obvious theorem:

Concurrence of altitudes. *In any triangle, the perpendiculars from the vertices to opposite sides (the* altitudes*) have a common point.*

To prove this theorem, take $\mathbf{0}$ at the intersection of two altitudes, say those through the vertices $\mathbf{u}$ and $\mathbf{v}$ (Figure 4.8). Then it remains to show that the line from $\mathbf{0}$ to the third vertex $\mathbf{w}$ is perpendicular to the side $\mathbf{v}-\mathbf{u}$.

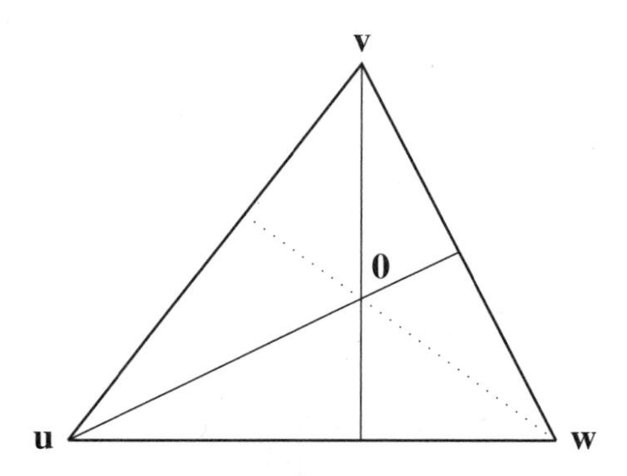

Figure 4.8: Altitudes of a triangle

Because $\mathbf{u}$ is perpendicular to the opposite side $\mathbf{w}-\mathbf{v}$, we have

$$\mathbf{u}\cdot(\mathbf{w}-\mathbf{v})=\mathbf{0}, \quad \text{that is,} \quad \mathbf{u}\cdot\mathbf{w}-\mathbf{u}\cdot\mathbf{v}=\mathbf{0}.$$

Because $\mathbf{v}$ is perpendicular to the opposite side $\mathbf{u}-\mathbf{w}$, we have

$$\mathbf{v}\cdot(\mathbf{u}-\mathbf{w})=\mathbf{0}, \quad \text{that is,} \quad \mathbf{v}\cdot\mathbf{u}-\mathbf{v}\cdot\mathbf{w}=\mathbf{0}.$$

Adding these two equations, and bearing in mind that $\mathbf{u}\cdot\mathbf{v}=\mathbf{v}\cdot\mathbf{u}$, we get

$$\mathbf{u}\cdot\mathbf{w}-\mathbf{v}\cdot\mathbf{w}=\mathbf{0}, \quad \text{that is,} \quad \mathbf{w}\cdot(\mathbf{v}-\mathbf{u})=\mathbf{0}.$$

Thus, $\mathbf{w}$ is perpendicular to $\mathbf{v}-\mathbf{u}$, as required. □

Exercises

The inner product criterion for directions to be perpendicular, namely that their inner product is zero, gives a neat way to prove the theorem in Exercise 2.2.2 about the diagonals of a rhombus.

4.4.1 Suppose that a parallelogram has vertices at $\mathbf{0}$, $\mathbf{u}$, $\mathbf{v}$, and $\mathbf{u}+\mathbf{v}$. Show that its diagonals have directions $\mathbf{u}+\mathbf{v}$ and $\mathbf{u}-\mathbf{v}$.

4.4.2 Deduce from Exercise 4.4.1 that the inner product of these directions is $|\mathbf{u}|^2-|\mathbf{v}|^2$, and explain why this is zero for a rhombus.

The inner product also gives a concise way to show that the equidistant line of two points is the perpendicular bisector of the line connecting them (thus proving more than we did in Section 3.3).

4.4.3 The condition for $\mathbf{w}$ to be equidistant from $\mathbf{u}$ and $\mathbf{v}$ is

$$(\mathbf{w}-\mathbf{u})\cdot(\mathbf{w}-\mathbf{u}) = (\mathbf{w}-\mathbf{v})\cdot(\mathbf{w}-\mathbf{v}).$$

Explain why, and show that this condition is equivalent to

$$|\mathbf{u}|^2 - 2\mathbf{w}\cdot\mathbf{u} = |\mathbf{v}|^2 - 2\mathbf{w}\cdot\mathbf{v}.$$

4.4.4 Show that the condition found in Exercise 4.4.3 is equivalent to

$$\left(\mathbf{w}-\frac{\mathbf{u}+\mathbf{v}}{2}\right)\cdot(\mathbf{u}-\mathbf{v}) = 0,$$

and explain why this says that $\mathbf{w}$ is on the perpendicular bisector of the line from $\mathbf{u}$ to $\mathbf{v}$.

Having established that the line equidistant from $\mathbf{u}$ and $\mathbf{v}$ is the perpendicular bisector, we conclude that the perpendicular bisectors of the sides of a triangle are concurrent—because this is obviously true of the equidistant lines of its vertices.

4.5 Inner product and cosine

The inner product of vectors $\mathbf{u}$ and $\mathbf{v}$ depends not only on their lengths $|\mathbf{u}|$ and $|\mathbf{v}|$ but also on the angle θ between them. The simplest way to express its dependence on angle is with the help of the *cosine* function. We write the cosine as a function of angle θ, $\cos\theta$. But, as usual, we avoid measuring angles and instead define $\cos\theta$ as the ratio of sides of a right-angled triangle. For simplicity, we assume that the triangle has vertices $\mathbf{0}$, $\mathbf{u}$, and $\mathbf{v}$ as shown in Figure 4.9.

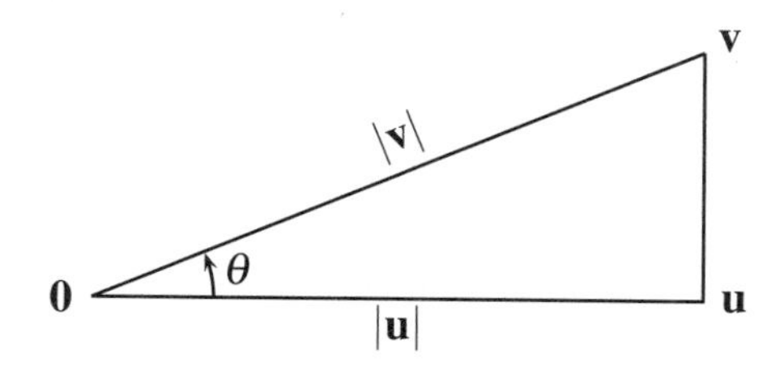

Figure 4.9: Cosine as a ratio of lengths

Then the side $\mathbf{v}$ is the hypotenuse, θ is the angle between the side $\mathbf{u}$ and the hypotenuse, and its cosine is defined by

$$\cos\theta = \frac{|\mathbf{u}|}{|\mathbf{v}|}.$$

We can now use the inner product criterion for perpendicularity to derive the following formula for inner product.

Inner product formula. *If θ is the angle between vectors* **u** *and* **v**, *then*

$$\mathbf{u}\cdot\mathbf{v} = |\mathbf{u}||\mathbf{v}|\cos\theta.$$

This formula follows because the side $\mathbf{v}-\mathbf{u}$ of the triangle is perpendicular to side $\mathbf{u}$; hence,

$$0 = \mathbf{u}\cdot(\mathbf{v}-\mathbf{u}) = \mathbf{u}\cdot\mathbf{v} - \mathbf{u}\cdot\mathbf{u}.$$

Therefore, $\mathbf{u}\cdot\mathbf{v} = \mathbf{u}\cdot\mathbf{u} = |\mathbf{u}|^2 = |\mathbf{u}||\mathbf{v}|\frac{|\mathbf{u}|}{|\mathbf{v}|} = |\mathbf{u}||\mathbf{v}|\cos\theta$. □

This formula gives a convenient way to calculate the angle (or at least its cosine) between any two lines, because we know from Section 4.4 how to calculate $|\mathbf{u}|$ and $|\mathbf{v}|$. It also gives us the "cosine rule" of trigonometry directly from the calculation of $(\mathbf{u}-\mathbf{v})\cdot(\mathbf{u}-\mathbf{v})$.

Cosine rule. *In any triangle, with sides* **u**, **v**, *and* $\mathbf{u}-\mathbf{v}$, *and angle θ opposite to the side* $\mathbf{u}-\mathbf{v}$,

$$|\mathbf{u}-\mathbf{v}|^2 = |\mathbf{u}|^2 + |\mathbf{v}|^2 - 2|\mathbf{u}||\mathbf{v}|\cos\theta.$$

Figure 4.10 shows the triangle and the relevant sides and angle, but the proof is a purely algebraic consequence of the inner product formula.

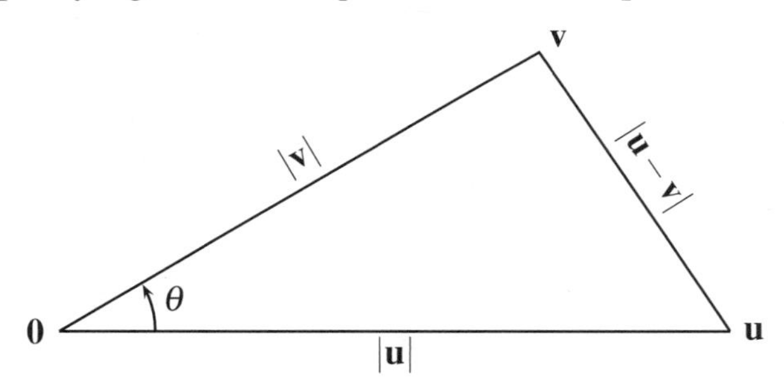

Figure 4.10: Quantities mentioned in the cosine rule

The algebra is simply the following:

$$\begin{aligned}|\mathbf{u}-\mathbf{v}|^2 &= (\mathbf{u}-\mathbf{v})\cdot(\mathbf{u}-\mathbf{v})\\ &= \mathbf{u}\cdot\mathbf{u} - 2\mathbf{u}\cdot\mathbf{v} + \mathbf{v}\cdot\mathbf{v}\\ &= |\mathbf{u}|^2 + |\mathbf{v}|^2 - 2\mathbf{u}\cdot\mathbf{v}\\ &= |\mathbf{u}|^2 + |\mathbf{v}|^2 - 2|\mathbf{u}||\mathbf{v}|\cos\theta.\end{aligned}$$

□

A nice way to close this circle of ideas is to consider the special case in which $\mathbf{u}$ and $\mathbf{v}$ are the sides of a right-angled triangle and $\mathbf{u}-\mathbf{v}$ is the hypotenuse. In this case, $\mathbf{u}$ is perpendicular to $\mathbf{v}$, so $\mathbf{u}\cdot\mathbf{v}=0$, and the cosine rule becomes

$$\text{hypotenuse}^2 = |\mathbf{u}-\mathbf{v}|^2 = |\mathbf{u}|^2 + |\mathbf{v}|^2$$

—which is the Pythagorean theorem. This result should not be a surprise, however, because we have already seen how the Pythagorean theorem is built into the definition of distance in $\mathbb{R}^2$ and hence into the inner product.

Exercises

The Pythagorean theorem can also be proved directly, by choosing $\mathbf{0}$ at the right angle of a right-angled triangle whose other two vertices are $\mathbf{u}$ and $\mathbf{v}$.

4.5.1 Show that $|\mathbf{v}-\mathbf{u}|^2 = |\mathbf{u}|^2 + |\mathbf{v}|^2$ under these conditions, and explain why this is the Pythagorean theorem.

While on the subject of right-angled triangles, we mention a useful formula for studying them.

4.5.2 Show that $(\mathbf{v}+\mathbf{u})\cdot(\mathbf{v}-\mathbf{u}) = |\mathbf{v}|^2 - |\mathbf{u}|^2$.

This formula gives a neat proof of the theorem from Section 2.7 about the angle in a semicircle. Take a circle with center $\mathbf{0}$ and a diameter with ends $\mathbf{u}$ and $-\mathbf{u}$ as shown in Figure 4.11. Also, let $\mathbf{v}$ be any other point on the circle.

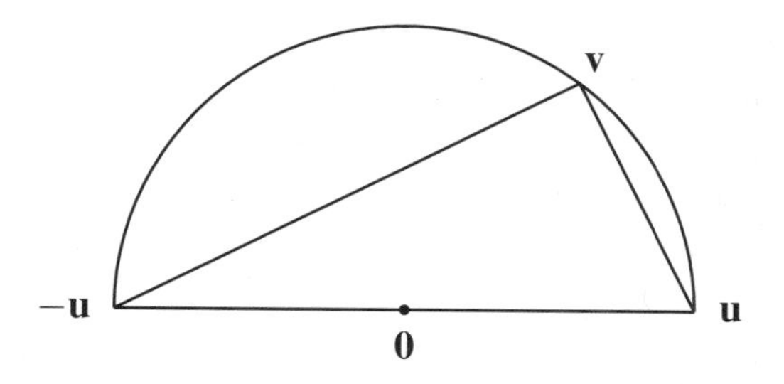

Figure 4.11: Points on a semicircle

4.5.3 Show that the sides of the triangle meeting at $\mathbf{v}$ have directions $\mathbf{v}+\mathbf{u}$ and $\mathbf{v}-\mathbf{u}$ and hence show that they are perpendicular.

4.6 The triangle inequality

In vector geometry, the triangle inequality $|\mathbf{u}+\mathbf{v}| \leq |\mathbf{u}|+|\mathbf{v}|$ of Exercises 3.3.1 to 3.3.3 is usually derived from the fact that

$$|\mathbf{u}\cdot\mathbf{v}| \leq |\mathbf{u}||\mathbf{v}|.$$

This result, known as the *Cauchy–Schwarz inequality*, follows easily from the formula in the previous section. The inner product formula says

$$\mathbf{u}\cdot\mathbf{v} = |\mathbf{u}||\mathbf{v}|\cos\theta,$$

and therefore,

$$\begin{aligned} |\mathbf{u}\cdot\mathbf{v}| &\leq |\mathbf{u}||\mathbf{v}||\cos\theta| \\ &\leq |\mathbf{u}||\mathbf{v}| \quad \text{because} \quad |\cos\theta| \leq \mathbf{1}. \end{aligned}$$

Now, to get the triangle inequality, it suffices to show that $|\mathbf{u}+\mathbf{v}|^2 \leq (|\mathbf{u}|+|\mathbf{v}|)^{\mathbf{2}}$, which we do as follows:

$$\begin{aligned} |\mathbf{u}+\mathbf{v}|^2 &= (\mathbf{u}+\mathbf{v})\cdot(\mathbf{u}+\mathbf{v}) \\ &= |\mathbf{u}|^{\mathbf{2}} + \mathbf{2u}\cdot\mathbf{v} + |\mathbf{v}|^{\mathbf{2}} \quad \text{because } \mathbf{u}\cdot\mathbf{u} = |\mathbf{u}|^{\mathbf{2}} \text{ and } \mathbf{v}\cdot\mathbf{v} = |\mathbf{v}|^{\mathbf{2}} \\ &\leq |\mathbf{u}|^{\mathbf{2}} + \mathbf{2}|\mathbf{u}||\mathbf{v}| + |\mathbf{v}|^{\mathbf{2}} \quad \text{by Cauchy–Schwarz} \\ &= (|\mathbf{u}|+|\mathbf{v}|)^{\mathbf{2}} \end{aligned}$$

□

The reason for the fuss about the Cauchy–Schwarz inequality is that it holds in spaces more complicated than $\mathbb{R}^2$, with more complicated inner products. Because the triangle inequality follows from Cauchy–Schwarz, it too holds in these complicated spaces. We are mainly concerned with the geometry of the plane, so we do not need complicated spaces. However, it is worth saying a few words about $\mathbb{R}^n$, because linear algebra works just as well there as it does in $\mathbb{R}^2$.

Higher dimensional Euclidean spaces

$\mathbb{R}^n$ is the set of ordered n-tuples $(x_1,x_2,\ldots,x_n)$ of real numbers $x_1,x_2,\ldots,x_n$. These ordered n-tuples are called *n-dimensional vectors*. If $\mathbf{u}$ and $\mathbf{v}$ are in $\mathbb{R}^n$, then we define the *vector sum* $\mathbf{u}+\mathbf{v}$ by

$$\mathbf{u}+\mathbf{v} = (u_1+v_1, u_2+v_2, \ldots, u_n+v_n),$$

and the *scalar multiple* $a\mathbf{u}$ for a real number a by

$$a\mathbf{u} = (au_1, au_2, \ldots, au_n).$$

It is easy to check that $\mathbb{R}^n$ has the properties enumerated at the beginning of Section 4.1. Hence, $\mathbb{R}^n$ is a real vector space under the vector sum and scalar multiplication operations just described.

$\mathbb{R}^n$ becomes a *Euclidean space* when we give it the extra structure of an inner product with the properties enumerated in Section 4.4. These properties hold if we define the inner product $\mathbf{u} \cdot \mathbf{v}$ by

$$\mathbf{u} \cdot \mathbf{v} = u_1 v_1 + u_2 v_2 + \cdots + u_n v_n,$$

as is easy to check. This inner product enables us to define *distance* in $\mathbb{R}^n$ by the formula

$$|\mathbf{u}|^2 = \mathbf{u} \cdot \mathbf{u}$$

which gives the distance $|\mathbf{u}|$ of $\mathbf{u}$ from the origin. This result is compatible with the concept of distance in $\mathbb{R}^2$ or $\mathbb{R}^3$ given by the Pythagorean theorem. For example, the distance of (u_1, u_2, u_3) from $\mathbf{0}$ in $\mathbb{R}^3$ is

$$|\mathbf{u}| = \sqrt{u_1^2 + u_2^2 + u_3^2},$$

as Figure 4.12 shows.

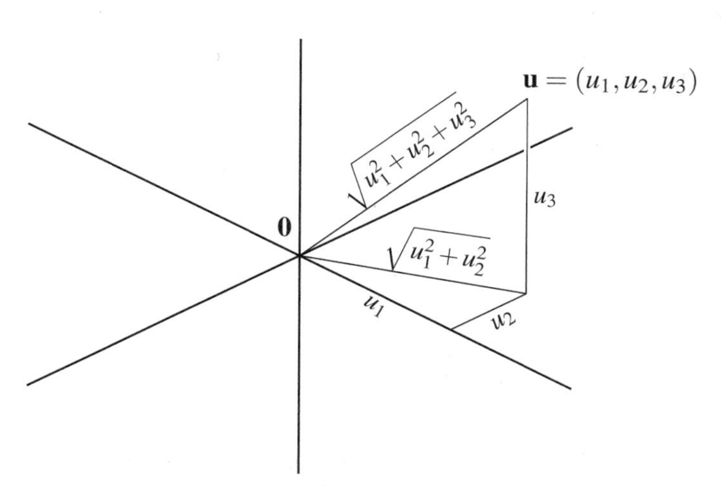

Figure 4.12: Distance in $\mathbb{R}^3$

- $\sqrt{u_1^2+u_2^2}$ is the distance from $\mathbf{0}$ of $(u_1,u_2,0)$ (the hypotenuse of a right-angled triangle with sides u_1 and u_2),

- $\sqrt{u_1^2+u_2^2+u_3^2}$ is the distance from $\mathbf{0}$ of (u_1,u_2,u_3) (the hypotenuse of a right-angled triangle with sides $\sqrt{u_1^2+u_2^2}$ and u_3).

All theorems proved in this chapter for vectors in the plane $\mathbb{R}^2$ hold in $\mathbb{R}^n$. This fact is clear if we take the plane in $\mathbb{R}^n$ to consist of vectors of the form $(x_1,x_2,0,\ldots,0)$, because such vectors behave exactly the same as vectors (x_1,x_2) in $\mathbb{R}^2$. But in fact *any* given plane in $\mathbb{R}^n$ behaves the same as the special plane of vectors $(x_1,x_2,0,\ldots,0)$. We skip the details, but it can be proved by constructing an isometry of $\mathbb{R}^n$ mapping the given plane onto the special plane. As in $\mathbb{R}^2$, any isometry is a product of reflections. In $\mathbb{R}^n$, at most $n+1$ reflections are required, and the proof is similar to the one given in Section 3.7.

Exercises

A proof of Cauchy–Schwarz using only general properties of the inner product can be obtained by an algebraic trick with quadratic equations. The general properties involved are the four listed at the beginning of Section 4.4 and the assumption that $\mathbf{w}\cdot\mathbf{w}=|\mathbf{w}|^2\geq 0$ for any vector $\mathbf{w}$ (an inner product with the latter property is called *positive definite*).

4.6.1 The Euclidean inner product for $\mathbb{R}^n$ defined above is positive definite. Why?

4.6.2 For any real number x, and any vectors $\mathbf{u}$ and $\mathbf{v}$, show that

$$(\mathbf{u}+x\mathbf{v})\cdot(\mathbf{u}+x\mathbf{v})=|\mathbf{u}|^2+2x(\mathbf{u}\cdot\mathbf{v})+x^2|\mathbf{v}|^2,$$

and hence that $|\mathbf{u}|^2+2x(\mathbf{u}\cdot\mathbf{v})+x^2|\mathbf{v}|^2\geq 0$ for any real number x.

4.6.3 If A, B, and C are real numbers and $A+Bx+Cx^2\geq 0$ for any real number x, explain why $B^2-4AC\leq 0$.

4.6.4 By applying Exercise 4.6.3 to the inequality $|\mathbf{u}|^2+2x(\mathbf{u}\cdot\mathbf{v})+x^2|\mathbf{v}|^2\geq 0$, show that

$$(\mathbf{u}\cdot\mathbf{v})^2\leq|\mathbf{u}|^2|\mathbf{v}|^2,\quad\text{and hence}\quad|\mathbf{u}\cdot\mathbf{v}|\leq|\mathbf{u}||\mathbf{v}|.$$

4.7 Rotations, matrices, and complex numbers

Rotation matrices

In Section 3.6, we defined a rotation of $\mathbb{R}^2$ as a function $r_{c,s}$, where c and s are two real numbers such that $c^2+s^2=1$. We described $r_{c,s}$ as the function that sends (x,y) to $(cx-sy, sx+cy)$, but it is also described by the *matrix of coefficients* of x and y, namely

$$\begin{pmatrix} c & -s \\ s & c \end{pmatrix}, \qquad \text{where } c=\cos\theta \text{ and } s=\sin\theta.$$

Because most readers will already have seen matrices, it may be useful to translate some previous statements about functions into matrix language, where they may be more familiar. (Readers not yet familiar with matrices will find an introduction in Section 7.2.)

Matrix notation allows us to rewrite $(x,y)\mapsto(cx-sy, sx+cy)$ as

$$\begin{pmatrix} c & -s \\ s & c \end{pmatrix}\begin{pmatrix} x \\ y \end{pmatrix}=\begin{pmatrix} cx-sy \\ sx+cy \end{pmatrix}$$

Thus, the function $r_{c,s}$ is applied to the variables x and y by multiplying the column vector $\begin{pmatrix} x \\ y \end{pmatrix}$ on the left by the matrix $\begin{pmatrix} c & -s \\ s & c \end{pmatrix}$. Functions are thereby separated from their variables, so they can be composed without the variables becoming involved—simply by multiplying matrices.

This idea gives proofs of the formulas for $\cos(\theta_1+\theta_2)$ and $\sin(\theta_1+\theta_2)$, similar to Exercises 3.5.3 and 3.5.4, but with the variables x and y filtered out:

- Rotation through angle θ_1 is given by the matrix $\begin{pmatrix} \cos\theta_1 & -\sin\theta_1 \\ \sin\theta_1 & \cos\theta_1 \end{pmatrix}$.
- Rotation through angle θ_2 is given by the matrix $\begin{pmatrix} \cos\theta_2 & -\sin\theta_2 \\ \sin\theta_2 & \cos\theta_2 \end{pmatrix}$.
- Hence, rotation through $\theta_1+\theta_2$ is given by the product of these two matrices. That is,

$$\begin{aligned}
&\begin{pmatrix} \cos(\theta_1+\theta_2) & -\sin(\theta_1+\theta_2) \\ \sin(\theta_1+\theta_2) & \cos(\theta_1+\theta_2) \end{pmatrix} \\
&= \begin{pmatrix} \cos\theta_1 & -\sin\theta_1 \\ \sin\theta_1 & \cos\theta_1 \end{pmatrix} \begin{pmatrix} \cos\theta_2 & -\sin\theta_2 \\ \sin\theta_2 & \cos\theta_2 \end{pmatrix} \\
&= \begin{pmatrix} \cos\theta_1\cos\theta_2 - \sin\theta_1\sin\theta_2 & -\cos\theta_1\sin\theta_2 - \sin\theta_1\cos\theta_2 \\ \cos\theta_1\sin\theta_2 + \sin\theta_1\cos\theta_2 & \cos\theta_1\cos\theta_2 - \sin\theta_1\sin\theta_2 \end{pmatrix}
\end{aligned}$$

by matrix multiplication.

- Finally, equating corresponding entries in the first and last matrices,

$$\cos(\theta_1+\theta_2) = \cos\theta_1\cos\theta_2 - \sin\theta_1\sin\theta_2,$$
$$\sin(\theta_1+\theta_2) = \cos\theta_1\sin\theta_2 + \sin\theta_1\cos\theta_2.$$

Complex numbers

One advantage of matrices, which we do not pursue here, is that they can be used to generalize the idea of rotation to any number of dimensions. But, for rotations of $\mathbb{R}^2$, there is a notation even more efficient than the rotation matrix

$$\begin{pmatrix} \cos\theta & -\sin\theta \\ \sin\theta & \cos\theta \end{pmatrix}.$$

It is the *complex number* $\cos\theta + i\sin\theta$, where $i = \sqrt{-1}$.

We represent the point $(x,y) \in \mathbb{R}^2$ by the complex number $z = x + iy$, and we rotate it through angle θ about O by *multiplying it* by $\cos\theta + i\sin\theta$. This procedure works because $i^2 = -1$, and therefore,

$$(\cos\theta + i\sin\theta)(x+iy) = x\cos\theta - y\sin\theta + i(x\sin\theta + y\cos\theta).$$

Thus, multiplication by $\cos\theta + i\sin\theta$ sends each point (x,y) to the point $(x\cos\theta - y\sin\theta, x\sin\theta + y\cos\theta)$, which is the result of rotating (x,y) about O through angle θ. Multiplying all points at once by $\cos\theta + i\sin\theta$, therefore, rotates the *whole plane* about O through angle θ.

It follows that multiplication by $(\cos\theta_1 + i\sin\theta_1)(\cos\theta_2 + i\sin\theta_2)$ rotates the plane through $\theta_1 + \theta_2$—the first factor rotates it through θ_1 and the second rotates it through θ_2—so it is the same as multiplication by

$\cos(\theta_1+\theta_2)+i\sin(\theta_1+\theta_2)$. Equating these two multipliers gives perhaps the ultimate proof of the formulas for $\cos(\theta_1+\theta_2)$ and $\sin(\theta_1+\theta_2)$:

$$\begin{aligned}\cos(\theta_1+\theta_2)+&i\sin(\theta_1+\theta_2)\\ &=(\cos\theta_1+i\sin\theta_1)(\cos\theta_2+i\sin\theta_2)\\ &=\cos\theta_1\cos\theta_2-\sin\theta_1\sin\theta_2+i(\cos\theta_1\sin\theta_2+\sin\theta_1\cos\theta_2)\\ &\quad\text{since } i^2=-1.\end{aligned}$$

Hence, equating real and imaginary parts,

$$\begin{aligned}\cos(\theta_1+\theta_2)&=\cos\theta_1\cos\theta_2-\sin\theta_1\sin\theta_2,\\ \sin(\theta_1+\theta_2)&=\cos\theta_1\sin\theta_2+\sin\theta_1\cos\theta_2.\end{aligned}$$

Exercises

The calculations above show that multiplication by $\cos\theta+i\sin\theta$ is rotation about O through angle θ because of the (seemingly accidental) property $i^2=-1$. In fact, any algebra of points in $\mathbb{R}^2$ that satisfies the same laws as the algebra of $\mathbb{R}$ automatically satisfies the condition $i^2=-1$, where i is the point $(0,1)$.

The following exercises show why. In particular, they reveal geometric consequences of the following algebraic laws:

$$|uv|=|u||v| \qquad \text{(multiplicative absolute value)}$$

$$u(v+w)=uv+uw \qquad \text{(distributive law)}$$

4.7.1 Given that $|x+iy|=\sqrt{x^2+y^2}$, explain why $|v-w|$ equals the distance between the complex numbers v and w.

4.7.2 Assuming the multiplicative absolute value and the distributive law (and, if necessary, any other algebraic laws satisfied by $\mathbb{R}$), show that

$$\text{distance between } uv \text{ and } uw=|u|\times\text{distance between } v \text{ and } w.$$

In other words, multiplying the plane $\mathbb{C}$ of complex numbers by a constant complex number u multiplies all distances by $|u|$.

4.7.3 Deduce from Exercises 4.7.1 and 4.7.2 that multiplication of $\mathbb{C}$ by a number u with $|u|=1$ is an isometry leaving O fixed.

4.7.4 Assuming that $u\neq 1$, and hence that $uz\neq z$ when $z\neq 0$, deduce from Exercise 4.7.3 that multiplication by $u\neq 1$ is a rotation.

These results explain why multiplication by u with $|u|=1$ is a rotation. To find the angle of rotation we assume that the point $(1,0)$ is the 1 of the algebra and observe where the rotation sends 1.

4.7.5 Explain why any u with $|u| = 1$ can be written in the form $\cos\theta + i\sin\theta$ for some angle θ, and conclude that multiplication by u rotates the point 1 (hence the whole plane) through angle θ.

It follows, in particular, that multiplication by $i = (0,1)$ sends $(1,0)$ to $(0,1)$ and hence rotates the plane through $\pi/2$. This result in turn implies $i^2 = -1$, because multiplication by i^2 then rotates the plane through π, which is also the effect of multiplication by -1.

4.8 Discussion

Because the geometric content of a vector space with an inner product is much the same as Euclidean geometry, it is interesting to see how many axioms it takes to describe a vector space. Remember from Section 2.9 that it takes 17 Hilbert axioms to describe the Euclidean plane, or 16 if we are willing to drop completeness of the line.

To define a vector space, we began in Section 4.1 with eight axioms for vector addition and scalar multiplication:

$$\mathbf{u}+\mathbf{v}=\mathbf{v}+\mathbf{u} \qquad 1\mathbf{u}=\mathbf{u}$$
$$\mathbf{u}+(\mathbf{v}+\mathbf{w})=(\mathbf{u}+\mathbf{v})+\mathbf{w} \qquad a(\mathbf{u}+\mathbf{v})=a\mathbf{u}+a\mathbf{v}$$
$$\mathbf{u}+\mathbf{0}=\mathbf{u} \qquad (a+b)\mathbf{u}=a\mathbf{u}+b\mathbf{u}$$
$$\mathbf{u}+(-\mathbf{u})=\mathbf{0} \qquad a(b\mathbf{u})=(ab)\mathbf{u}.$$

Then, in Section 4.4, we added three (or four, depending on how you count) axioms for the inner product:

$$\mathbf{u}\cdot\mathbf{v}=\mathbf{v}\cdot\mathbf{u},$$
$$\mathbf{u}\cdot(\mathbf{v}+\mathbf{w})=\mathbf{u}\cdot\mathbf{v}+\mathbf{u}\cdot\mathbf{w},$$
$$(a\mathbf{u})\cdot\mathbf{v}=\mathbf{u}\cdot(a\mathbf{v})=a(\mathbf{u}\cdot\mathbf{v}),$$

We also need relations among inner product, length, and angle—at a minimum the cosine formula,

$$\mathbf{u}\cdot\mathbf{v}=|\mathbf{u}||\mathbf{v}|\cos\theta,$$

so this is 12 or 13 axioms so far.

But we have also assumed that the scalars $a, b, \ldots$ are real numbers, so there remains the problem of writing down axioms for them. At the very least, one needs axioms saying that the scalars satisfy the ordinary rules of calculation, the so-called *field axioms* (this is usual when defining a vector space):

$$a+b=b+a, \qquad ab=ba \qquad \text{(commutative laws)}$$
$$a+(b+c)=(a+b)+c, \qquad a(bc)=(ab)c \qquad \text{(associative laws)}$$
$$a+0=a, \qquad a1=a \qquad \text{(identity laws)}$$
$$a+(-a)=0, \qquad aa^{-1}=1 \qquad \text{(inverse laws)}$$
$$a(b+c)=ab+ac \qquad \text{(distributive law)}$$

Thus, the usual definition of a vector space, with an inner product suitable for Euclidean geometry, takes more than 20 axioms! Admittedly, the field axioms and the vector space axioms are useful in many other parts of mathematics, whereas most of the Hilbert axioms seem meaningful only in geometry. And, by varying the inner product slightly, one can change the geometry of the vector space in interesting ways. For example, one can obtain the geometry of *Minkowski space* used in Einstein's special theory of relativity. To learn more about the vector space approach to geometry, see *Linear Algebra and Geometry, a Second Course* by I. Kaplansky and *Metric Affine Geometry* by E. Snapper and R. J. Troyer.

Still, one can dream of building geometry on a much simpler set of axioms. In Chapter 6, we will realize this dream with *projective geometry*, which we begin studying in Chapter 5.

5

Perspective

PREVIEW

Euclid's geometry concerns figures that can be drawn with straightedge and compass, even though many of its theorems are about straight lines alone. Are there any interesting figures that can be drawn with straightedge alone? Remember, the straightedge has no marks on it, so it is impossible to copy a length. Thus, with a straightedge alone, we cannot draw a square, an equilateral triangle, or any figure involving equal line segments. Yet there is something interesting we *can* draw: a *perspective view of a tiled floor*, such as the one shown in Figure 5.1.

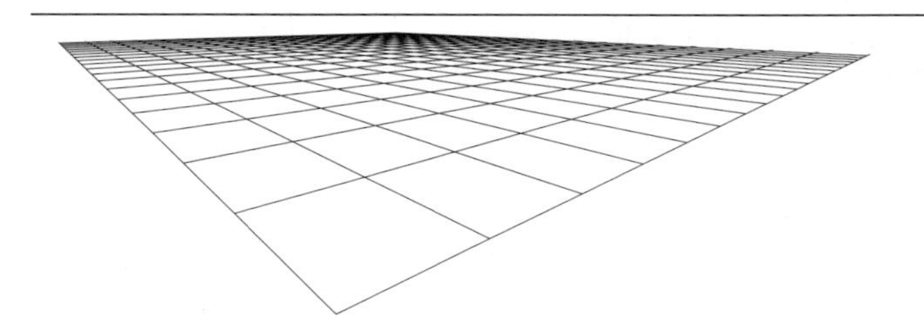

Figure 5.1: Perspective view of a tiled floor

This picture is interesting because it seems clear that all tiles in the view are of equal size. Thus, even though we cannot draw tiles that are actually equal, we can draw tiles that *look* equal.

We will explain how to solve the problem of drawing perspective views in Section 5.2. The solution takes us into a new form of geometry—a geometry of vision—called *projective geometry*.

5.1 Perspective drawing

Sometime in the 15th century, Italian artists discovered how to draw three-dimensional scenes in correct perspective. Figures 5.2 and 5.3 illustrate the great advance in realism this skill achieved, with pictures drawn before and after the discovery. The "before" picture, Figure 5.2, is a drawing I found in the book *Perspective in Perspective* by L. Wright. It is thought to date from the late 15th century, but it comes from England, where knowledge of perspective had evidently not reached at that time.

Figure 5.2: *The birth of St Edmund*, by an unknown artist

The "after" picture, Figure 5.3, is the 1514 engraving *Saint Jerome in his study*, by the great German artist Albrecht Dürer (1471–1528). Dürer made study tours of Italy in 1494 and 1505 and became a master of all aspects of drawing, including perspective.

The simplest test of perspective drawing is the depiction of a tiled floor. The picture in Figure 5.2 clearly fails this test. All the tiles are drawn as rectangles, which makes the floor look vertical. We know from experience that a horizontal rectangle does not *look* rectangular—its angles are not all right angles because its sides converge to a common point on the horizon, as in the tabletop in Dürer's engraving.

Figure 5.3: *St Jerome in his study*, by Albrecht Dürer

The Italians drew tiles by a method called the *costruzione legittima* (legitimate construction), first published by Leon Battista Alberti in 1436. The bottom edge of the picture coincides with a line of tile edges, and any other horizontal line is chosen as the horizon. Then lines drawn from equally spaced points on the bottom edge to a point on the horizon depict the parallel columns of tiles perpendicular to the bottom edge (Figure 5.4). Another horizontal line, near the bottom, completes the first row of tiles.

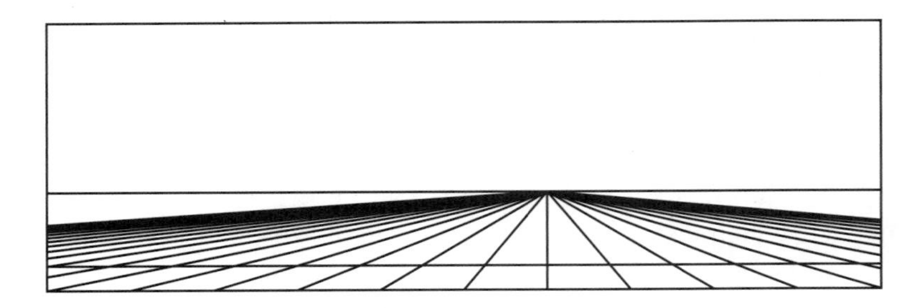

Figure 5.4: Beginning the *costruzione legittima*

The real problem comes next. How do we find the correct horizontal lines to depict the 2nd, 3rd, 4th, ... rows of tiles? The answer is surprisingly simple: Draw the *diagonal* of any tile in the bottom row (shown in gray in Figure 5.5). The diagonal necessarily crosses successive columns at the corners of tiles in the 2nd, 3rd, 4th, ... rows; hence, these rows can be constructed by drawing horizontal lines at the successive crossings. Voilà!

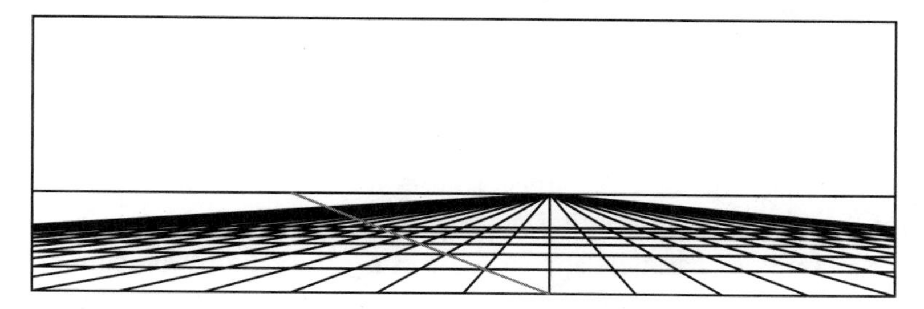

Figure 5.5: Completing the *costruzione legittima*

Exercises

Suppose that the floor has rows of tiles crossing the x-axis at $x = 0, 1, 2, 3, \ldots$, and that the artist copies the view of the floor onto a vertical transparent screen through the y-axis, keeping a fixed eye position at the point $(-1, 1)$. Then the perspective view of the points $x = 0, 1, 2, 3, \ldots$ will be the series of points on the y-axis shown in Figure 5.6.

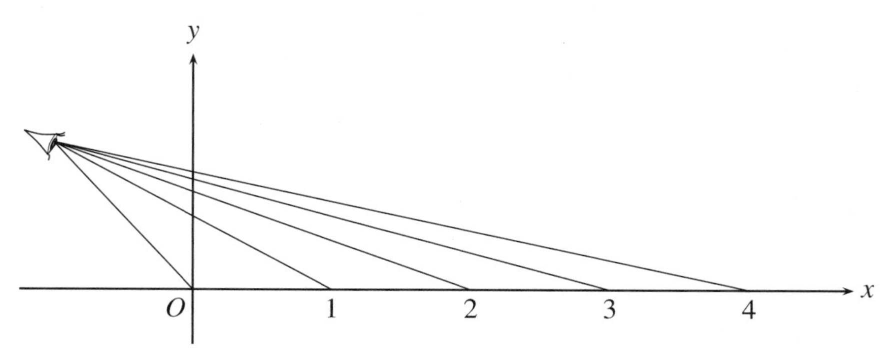

Figure 5.6: Perspective view of equally spaced points

5.1.1 Show that the line from $(-1, 1)$ to $(n, 0)$ crosses the y-axis at $y = \frac{n}{n+1}$. Hence, the perspective images of the points $x = 0, 1, 2, 3, \ldots$ are the points $y = 0, \frac{1}{2}, \frac{2}{3}, \frac{3}{4}, \ldots$.

If each of the points $0, 1, 2, 3, \ldots$ is sent to the next, then each of their perspective images $y = 0, \frac{1}{2}, \frac{2}{3}, \frac{3}{4}, \ldots$ is sent to the next.

5.1.2 Show that the function $f(y) = \frac{1}{2-y}$ effects this move.

5.1.3 Which point on the y-axis is not moved by the function $f(y) = \frac{1}{2-y}$, and what is the geometric significance of this point?

5.2 Drawing with straightedge alone

The *construzione legittima* takes advantage of something that is visually obvious but mathematically mysterious—the fact that parallel lines generally do not look parallel, but appear to meet on the horizon. The point where a family of parallels appear to meet is called their "vanishing point" by artists, and their *point at infinity* by mathematicians. The horizon itself, which consists of all the points at infinity, is called the *line at infinity*.

However, the *costruzione legittima* does not take full advantage of points at infinity. It involves some parallels that are really *drawn* parallel, so we need both straightedge and compass as used in Chapter 1. The construction also needs measurement to lay out the equally spaced points on the bottom line of the picture, and this again requires a compass. Thus, the *costruzione legittima* is a Euclidean construction at heart, requiring both a straightedge and a compass.

Is it possible to draw a perspective view of a tiled floor with a straightedge alone? Absolutely! All one needs to get started is the horizon and a tile placed obliquely. The tile is created by the two pairs of parallel lines, which are simply pairs that meet on the horizon (Figure 5.7).

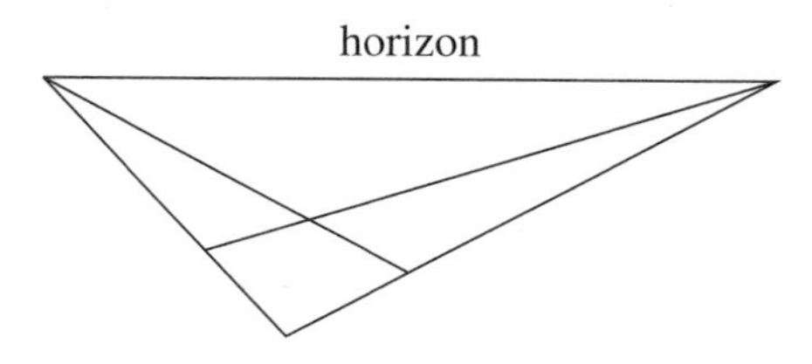

Figure 5.7: The first tile

We then draw the diagonal of this tile and extend it to the horizon, obtaining the point at infinity of all diagonals parallel to this first one. This step allows us to draw two more diagonals, of tiles adjacent to the first one.

These diagonals give us the remaining sides of the adjacent tiles, and we can then repeat the process. The first few steps are shown in Figure 5.8. Figure 5.1 at the beginning of the chapter is the result of carrying out many steps (and deleting the construction lines).

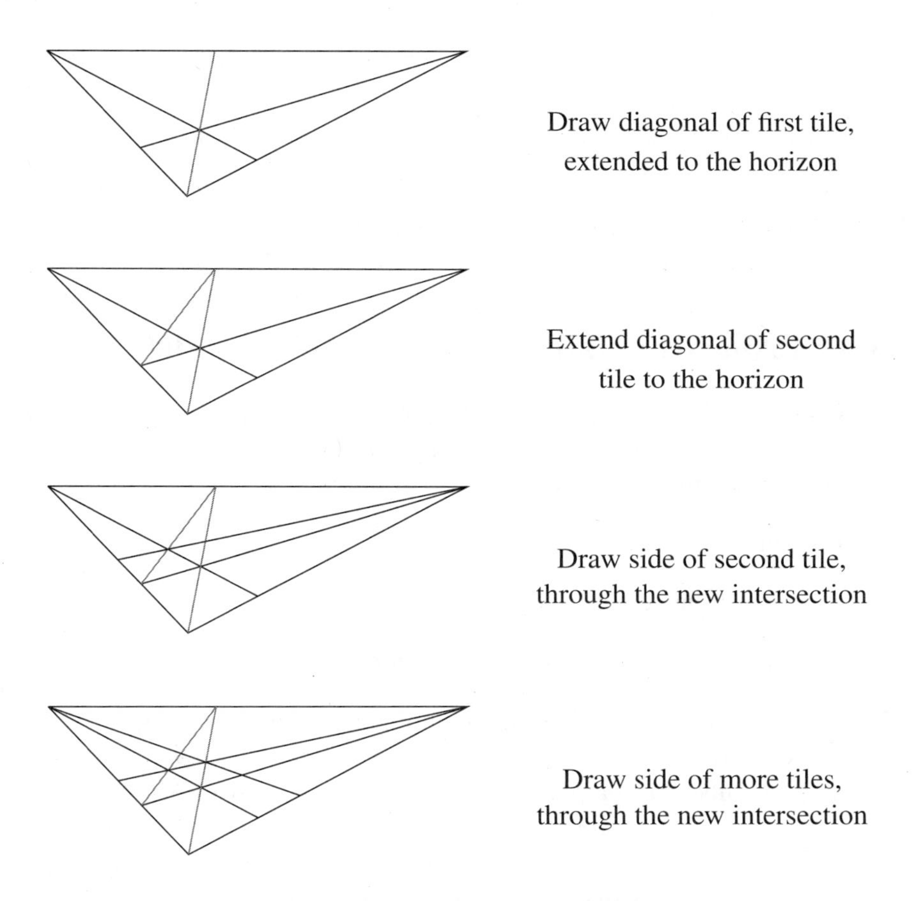

Figure 5.8: Constructing the tiled floor

This construction is easy and fun to do, and we urge the reader to get a straightedge and try it. Also try the constructions suggested in the exercises, which create pictures of floors with differently shaped tiles.

Exercises

Consider the triangular tile shown shaded in Figure 5.9. Notice that this triangle could be half of the quadrangular tile shown in Figure 5.7 (this is a hint).

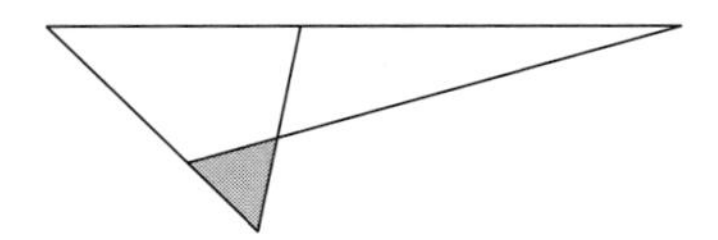

Figure 5.9: A triangular tile

5.2.1 Draw a perspective view of the plane filled with many copies of this tile.

5.2.2 Also, by deleting some lines in your solution to Exercise 5.2.1, create a perspective view of the plane filled with congruent hexagons.

5.3 Projective plane axioms and their models

Drawing a tiled floor with straightedge alone requires a "horizon"—a line at infinity. Apart from this requirement, the construction works because certain things remain the same in any view of the plane:

- straight lines remain straight
- intersections remain intersections
- parallel lines remain parallel or meet on the horizon.

Now parallel lines *always* meet on the horizon if you point yourself in the right direction, so if we could look in all directions at once we would see that any two lines have a point in common. This idea leads us to believe in a structure called a *projective plane*, containing objects called *"points"* and *"lines"* satisfying the following axioms. We write "points" and "lines" in quotes because they may not be the same as ordinary points and lines.

Axioms for a projective plane

1. Any two "points" are contained in a unique "line."
2. Any two "lines" contain a unique "point."
3. There exist four "points", no three of which are in a "line."

Notice that these are axioms about *incidence*: They involve only meetings between "points" and "lines," not things such as length or angle. Some of Euclid's and Hilbert's axioms are of this kind, but not many.

Axiom 1 is essentially Euclid's first axiom for the construction of lines. Axiom 2 says that there are no exceptional pairs of lines that do not meet. We can define "parallels" to be lines that meet on a line called the "horizon," but this does not single out a special class of lines—in a projective plane, the "horizon" behaves the same as any other line. Axiom 3 says that a projective plane has "enough points to be interesting." We can think of the four points as the four vertices of a quadrilateral, from which one may generate the complicated structure seen in the pictures of a tiled floor at the beginning of this chapter.

The real projective plane

If there is such a thing as a projective plane, it should certainly satisfy these axioms. But does *anything* satisfy them? After all, we humans can never see all of the horizon at once, so perhaps it is inconsistent to suppose that all parallels meet. These doubts are dispelled by the following *model*, or *interpretation*, of the axioms for a projective plane. The model is called the *real projective plane* $\mathbb{RP}^2$, and it gives a mathematical meaning to the terms "point," "line," and "plane" that makes all the axioms true.

Take "points" to be lines through O in $\mathbb{R}^3$, "lines" to be planes through O in $\mathbb{R}^3$, and the "plane" to be the set of all lines through O in $\mathbb{R}^3$. Then

1. Any two "points" are contained in a unique "line" because two given lines through O lie in a unique plane through O.

2. Any two "lines" contain a unique "point" because any two planes through O meet in a unique line through O.

3. There are four different "points," no three of which are in a "line": for example, the lines from O to the four points $(1,0,0)$, $(0,1,0)$, $(0,0,1)$, and $(1,1,1)$, because no three of these lines lie in the same plane through O.

The last claim is perhaps a little hard to grasp by visualization, but it can be checked algebraically because any plane through O has an equation of the form

$$ax + by + cz = 0 \quad \text{for some real numbers} \quad a, b, c.$$

If, say, $(1,0,0)$ and $(0,1,0)$ are on this plane, then we find by substituting these values of x,y,z in the equation that

$$a=0 \quad \text{and} \quad b=0, \quad \text{hence the plane is } z=0.$$

But $(0,0,1)$ and $(1,1,1)$ do *not* lie on the plane $z=0$. It can be checked similarly that the plane through any two of the points does not contain the other two.

It is no fluke that lines and planes through O in $\mathbb{R}^3$ behave as we want "points" and "lines" of a projective plane to behave, because they capture the idea of *viewing with an all-seeing eye.* The point O is the position of the eye, and the lines through O connect the eye to points in the plane. Consider how the eye sees the plane $z=-1$, for example (Figure 5.10).

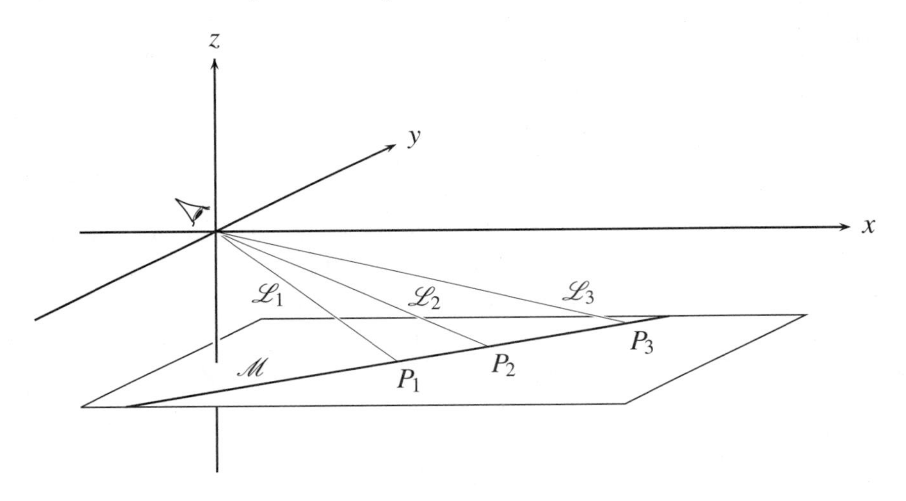

Figure 5.10: Viewing a plane from O

Points $P_1,P_2,P_3,\ldots$ in the plane $z=-1$ are joined to the eye by lines $\mathscr{L}_1,\mathscr{L}_2,\mathscr{L}_3,\ldots$ through O, and as the point P_n tends to infinity, the line $\mathscr{L}_n$ tends toward the horizontal. Therefore, it is natural to call the horizontal lines through O the "points at infinity" of the plane $z=-1$, and to call the plane of all horizontal lines through O the "horizon" or "line at infinity" of the plane $z=-1$.

Unlike the lines $\mathscr{L}_1,\mathscr{L}_2,\mathscr{L}_3,\ldots$, corresponding to points $P_1,P_2,P_3,\ldots$ of the *Euclidean plane* $z=-1$, horizontal lines through O have no counterparts in the Euclidean plane: They *extend* the Euclidean plane to a *projective plane*. However, the extension arises in a natural way. Once we replace the points $P_1, P_2, P_3, \dots$ by lines in space, we realize that there are extra lines (the horizontal lines) corresponding to the points on the horizon.

This model of the projective plane nicely captures our intuitive idea of points at infinity, but it also makes the idea clearer. We can see, for example, why it is proper for each line to have only one point at infinity, not two: because the lines $\mathscr{L}$ connecting O to points P along a line $\mathscr{M}$ in the plane $z = -1$ tend toward the *same* horizontal line as P tends to infinity in either direction (namely, the parallel to $\mathscr{M}$ through O).

It is hard to find a surface that behaves like $\mathbb{RP}^2$, but it is easy to find a curve that behaves like any "line" in it, a so-called *real projective line*. Figure 5.11 shows how. The "points" in a "line" of $\mathbb{RP}^2$, namely the lines through O in some plane through O, correspond to points of a circle through O. Each point $P \neq O$ on the circle corresponds to the line through O and P, and the point O itself corresponds to the tangent line at O.

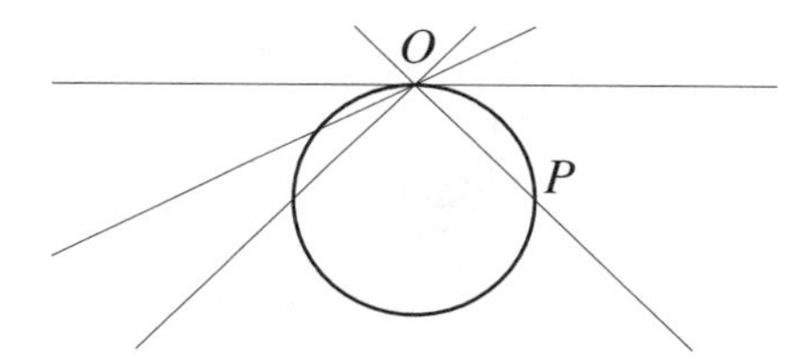

Figure 5.11: Modeling a projective line by a circle

Exercises

To gain more familiarity with calculations in $\mathbb{R}^3$, let us pursue the example of four "points" given above.

5.3.1 Find the plane $ax + by + cz = 0$ through the points $(0,0,1)$ and $(1,1,1)$, and check that it does not contain the points $(1,0,0)$ and $(0,1,0)$.

5.3.2 Show that $\mathbb{RP}^2$ has four "lines," no three of which have a common "point."

Not only does $\mathbb{RP}^2$ contain four "lines," no three of which have a "point" in common; the same is true of *any* projective plane, because this property follows *from the projective plane axioms alone*.

5.3.3 Suppose that A, B, C, D are four "points" in a projective plane, no three of which are in a "line." Consider the "lines" AB, BC, CD, DA. Show that if AB and BC have a common point E, then $E = B$.

5.3.4 Deduce from Exercise 5.3.3 that the three lines AB, BC, CD have no common point, and that the same is true of any three of the lines AB, BC, CD, DA.

5.4 Homogeneous coordinates

Because "points" and "lines" of $\mathbb{RP}^2$ are lines and planes through O in $\mathbb{R}^3$, they are easily handled by methods of linear algebra. A line through O is determined by any point $(x,y,z) \neq O$, and it consists of the points (tx,ty,tz), where t runs through all real numbers. Thus, a "point" is not given by a single triple (x,y,z), but rather by any of its nonzero multiples (tx,ty,tz). These triples are called the *homogeneous coordinates* of the "point."

A plane through O has a linear equation of the form $ax+by+cz=0$, called a *homogeneous equation*. The same plane is given by the equation $tax+tby+tcz=0$ for any nonzero t. Thus, a "line" is likewise not given by a single triple (a,b,c), but by the set of all its nonzero multiples (ta,tb,tc).

If (x_1,y_1,z_1) and (x_2,y_2,z_2) lie on different lines through O, then it is geometrically obvious that they lie in a unique plane $ax+by+cz=0$. The coordinates (a,b,c) of this plane can be found by solving the two equations

$$\begin{aligned} ax_1+by_1+cz_1 &= 0, \\ ax_2+by_2+cz_2 &= 0, \end{aligned}$$

for a, b, and c. Because there are more unknowns than equations, there is not a single solution triple but a whole space of them—in this case, a set of multiples (ta,tb,tc), all representing the same homogeneous equation.

This is the algebraic reason why two "points" lie on a unique "line" in $\mathbb{RP}^2$. There is a similar reason why two "lines" have a unique "point" in common. Two "lines" are given by two equations

$$\begin{aligned} a_1x+b_1y+c_1z &= 0, \\ a_2x+b_2y+c_2z &= 0, \end{aligned}$$

and we find their common "point" by solving these equations for x, y, and z. This problem is the same as above, but with the roles of a,b,c exchanged with those of x,y,z. The solution in this case is a set of multiples (tx,ty,tz) representing the homogeneous coordinates of the common "point."

The practicalities of finding the "line" through two "points" or the "point" common to two "lines" are explored in the next exercise set. But first I want to make a theoretical point. *It makes no algebraic difference if the coordinates of "points" and "lines" are complex numbers.* We can define a *complex projective plane* $\mathbb{CP}^2$, each "point" of which is a set of triples of the form (tx,ty,tz), where x,y,z are particular complex numbers and t runs through all complex numbers. It remains true that any two "points" lie on a unique "line" and any two "lines" have unique common point, simply because the algebraic properties of complex linear equations are exactly the same as those of real linear equations. Similarly, one can show there are four "points," no three of which are in a "line" of $\mathbb{CP}^2$.

Thus, there is more than one model of the projective plane axioms. Later we shall look at other models, which enable us to see that certain properties of $\mathbb{RP}^2$ are not properties of all projective planes and hence do not follow from the projective plane axioms.

Projective space

It is easy to generalize homogeneous coordinates to *quadruples* (w,x,y,z) and hence to define the three-dimensional *real projective space* $\mathbb{RP}^3$. It has "points," "lines," and "planes" defined as follows (we use vector notation to shorten the definitions):

- A "point" is a line through O in $\mathbb{R}^4$, that is, a set of quadruples $t\mathbf{u}$, where $\mathbf{u} = (w,x,y,z)$ is a particular quadruple of real numbers and t runs through all real numbers.
- A "line" is a plane through O in $\mathbb{R}^4$, that is, a set $t_1\mathbf{u}_1 + t_2\mathbf{u}_2$ where $\mathbf{u}_1$ and $\mathbf{u}_2$ are linearly independent points of $\mathbb{R}^4$ and t_1 and t_2 run through all real numbers.
- A "plane" is a three-dimensional space through O in $\mathbb{R}^4$, that is, a set $t_1\mathbf{u}_1 + t_2\mathbf{u}_2 + t_3\mathbf{u}_3$, where $\mathbf{u}_1$, $\mathbf{u}_2$, and $\mathbf{u}_3$ are linearly independent points of $\mathbb{R}^4$ and t_1, t_2, and t_3 run through all real numbers.

Linear algebra then enables us to show various properties of the "points," "lines," and "planes" in $\mathbb{RP}^3$, such as:

1. Two "points" lie on a unique "line."
2. Three "points" not on a "line" lie on a unique "plane."
3. Two "planes" have unique "line" in common.
4. Three "planes" with no common "line" have one common "point."

These properties hold for any *three-dimensional projective space*, and $\mathbb{RP}^3$ is not the only one. There is also a complex projective space $\mathbb{CP}^3$, and many others. $\mathbb{RP}^3$ has an unexpected influence on the geometry of the sphere, as we will see in Section 7.8.

Exercises

5.4.1 Find the plane $ax+by+cz=0$ that contains the points $(1,2,3)$ and $(1,1,1)$.

5.4.2 Find the line of intersection of the planes $x+2y+3z=0$ and $x+y+z=0$.

5.4.3 You can write down the solution of Exercise 5.4.2 as soon as you have solved Exercise 5.4.1. Why?

5.5 Projection

The three-dimensional Euclidean space $\mathbb{R}^3$, in which the lines through O are the "points" of $\mathbb{RP}^2$ and the planes through O are the "lines" of $\mathbb{RP}^2$, also contains many other planes. Each plane $\mathscr{P}$ *not* passing through O can be regarded as a *perspective view* of the projective plane $\mathbb{RP}^2$, a view that contains all but one "line" of $\mathbb{RP}^2$.

Each point P of $\mathscr{P}$ corresponds to a line ("of sight") through O, and hence to a "point" of $\mathbb{RP}^2$. The only lines through O that do not meet $\mathscr{P}$ are those parallel to $\mathscr{P}$, and these make up the *line at infinity* or *horizon* of $\mathscr{P}$, as we have already seen in the case of the plane $z=-1$ in Section 5.3.

If $\mathscr{P}_1$ and $\mathscr{P}_2$ are any two planes not passing through O we can *project* $\mathscr{P}_1$ *to* $\mathscr{P}_2$ by sending each point P_1 in $\mathscr{P}_1$ to the point P_2 in $\mathscr{P}_2$ lying on the same line through O as P_1 (Figure 5.12). The geometry of $\mathbb{RP}^2$ is called "projective" because it encapsulates the geometry of a whole family of planes related by projection.

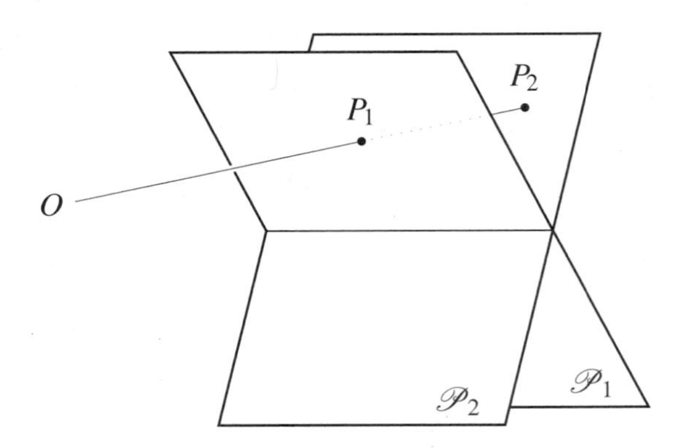

Figure 5.12: Projecting one plane to another

Projections of projective lines

Projection of one plane $\mathscr{P}_1$ onto another plane $\mathscr{P}_2$ produces an image of $\mathscr{P}_1$ that is generally distorted in some way. For example, a grid of squares on $\mathscr{P}_1$ may be mapped to a perspective view of the grid that looks like Figure 5.1. Nevertheless, straight lines remain straight under projection, so there are limits to the amount of distortion in the image. To better understand the nature and scope of projective distortion, in this subsection we analyze the mappings of the projective *line* obtainable by projection.

An effective way to see the distortion produced by projection of one line $\mathscr{L}_1$ onto another line $\mathscr{L}_2$ is to mark a series of equally spaced dots on $\mathscr{L}_1$ and the corresponding image dots on $\mathscr{L}_2$. You can think of the image dots as "shadows" of the dots on $\mathscr{L}_1$ cast by light rays from the point of projection P, except that we have projective lines through P, not rays, so it can seem as though the "shadow" on $\mathscr{L}_2$ comes ahead of the dot on $\mathscr{L}_1$. (See Figure 5.15, but bear in mind that a projective line is really circular, so it is always possible to pass through P, to a point on $\mathscr{L}_1$, then to a point on $\mathscr{L}_2$, in that order.)

In the simplest cases, where $\mathscr{L}_1$ and $\mathscr{L}_2$ are parallel, the image dots are also equally spaced. Figure 5.13 shows the case of *projection from a point at infinity*, where the lines from the dots on $\mathscr{L}_1$ to their images on $\mathscr{L}_2$ are parallel and hence the dots on $\mathscr{L}_1$ are simply translated a constant distance l. If we choose an origin on each line and use the same unit of length on each, then projection from infinity sends each x on $\mathscr{L}_1$ to $x+l$ on $\mathscr{L}_2$.

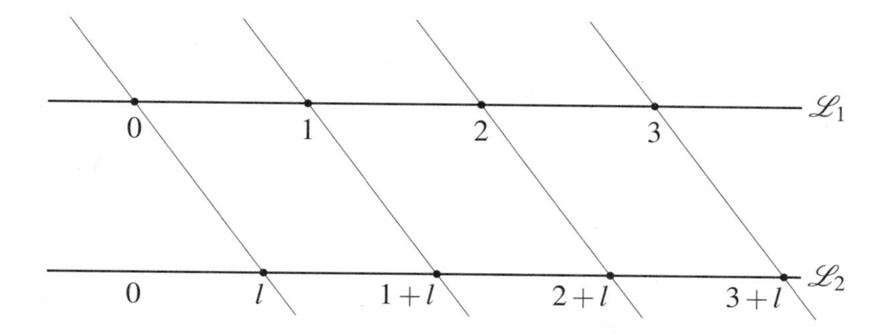

Figure 5.13: Projection from infinity

When $\mathscr{L}_1$ is projected from a finite point P, then the distance between dots is magnified by a constant factor $k \neq 0$. If we take P on a line through the zero points on $\mathscr{L}_1$ and $\mathscr{L}_2$, then the projection sends each x on $\mathscr{L}_1$ to kx on $\mathscr{L}_2$ (Figure 5.14). Note also that this projection sends x on $\mathscr{L}_2$ to x/k on $\mathscr{L}_1$, so the magnification factor can be *any* $k \neq 0$.

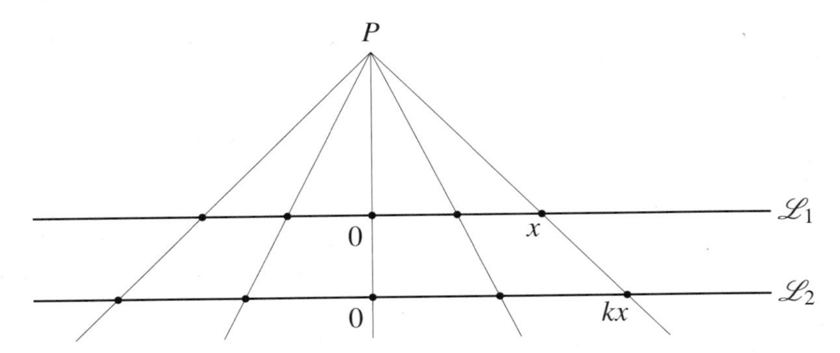

Figure 5.14: Projection from a finite point

When $\mathscr{L}_1$ and $\mathscr{L}_2$ are not parallel the distortion caused by projection is more extreme. Figure 5.15 shows how the spacing of dots changes when $\mathscr{L}_1$ is projected onto a perpendicular line $\mathscr{L}_2$ from a point O equidistant from both. Figure 5.16 is a closeup of the image line $\mathscr{L}_2$, showing how the image dots "converge" to a point corresponding to the horizontal line through O (which corresponds to the point at infinity on $\mathscr{L}_1$).

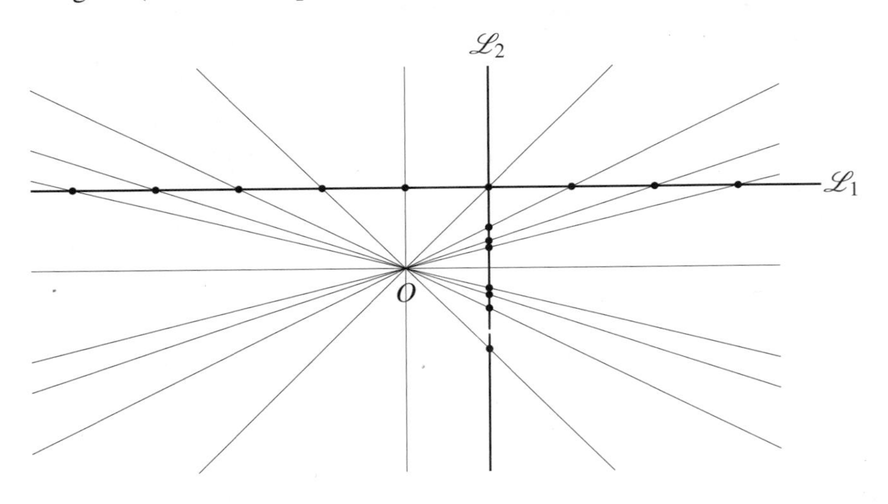

Figure 5.15: Example of projective distortion of the line

Figure 5.16: Closeup of the image line

We take $O = (0,0)$ as usual, and we suppose that $\mathscr{L}_1$ is parallel to the x-axis, that $\mathscr{L}_2$ is parallel to the y-axis, and that the dots on $\mathscr{L}_1$ are unit distance apart. Then the line from O to the dot at $x = n$ on $\mathscr{L}_1$ has slope $1/n$ and hence it meets the line $\mathscr{L}_2$ at $y = 1/n$. Thus the map from $\mathscr{L}_1$ to $\mathscr{L}_2$ is the function sending x to $y = 1/x$. This map exhibits the most extreme kind of distortion induced by projection, with the point at infinity on $\mathscr{L}_1$ sent to the point $y = 0$ on $\mathscr{L}_2$.

Any combination of these projections is therefore a combination of functions $1/x$, kx, and $x + l$, which are called *generating transformations*. The combinations of generating transformations are precisely the functions of the form

$$f(x) = \frac{ax+b}{cx+d}, \quad \text{where} \quad ad - bc \neq 0,$$

that we study in the next section.

Exercises

Before studying all these functions, it is useful to study the (simpler) subclass obtained by composing functions that send x to $x + l$ or kx (for $k \neq 0$). The latter functions obviously include any function of the form $f(x) = ax + b$ with $a \neq 0$, which is the result of multiplying by a, and then adding b.

5.5.1 If $f_1(x) = a_1 x + b_1$ with $a_1 \neq 0$ and $f_2(x) = a_2 x + b_2$ with $a_2 \neq 0$, show that

$$f_1(f_2(x)) = Ax + B, \quad \text{with} \quad A \neq 0,$$

and find the constants A and B.

5.5.2 Deduce from Exercise 5.5.1 that the result of composing any number of functions that send x to $x + l$ or kx (for $k \neq 0$) is a function of the form $f(x) = ax + b$ with $a \neq 0$.

We know that such functions represent combinations of certain projections from lines to parallel lines, but do they include *any* projection from a line to a parallel line?

5.5.3 Show that projection of a line, from any finite point P, onto a parallel line is represented by a function of the form $f(x) = ax + b$.

5.6 Linear fractional functions

The functions sending x to $1/x$, kx, and $x+l$ are among the functions called *linear fractional*, each of which has the form

$$f(x) = \frac{ax+b}{cx+d} \quad \text{where} \quad ad-bc \neq 0.$$

The condition $ad-bc \neq 0$ ensures that $f(x)$ is not constant. Constancy occurs only if $ax+b = \frac{a}{c}(cx+d)$; in which case, $ad-bc=0$ because $\frac{ad}{c} = b$.

By writing

$$\frac{ax+b}{cx+d} \quad \text{as} \quad \frac{ax+\frac{ad}{c}+b-\frac{ad}{c}}{cx+d} = \frac{\frac{a}{c}(cx+d)+\frac{1}{c}(bc-ad)}{cx+d}$$

we find that any linear fractional function with $c \neq 0$ may be written in the form

$$f(x) = \frac{a}{c} + \frac{bc-ad}{c(cx+d)}.$$

Such a function may therefore be composed from functions sending x to $1/x$, kx, and $x+l$—the functions that reciprocate, multiply by k, and add l—for various values of k and l:

- first multiply x by c,
- then add d,
- then multiply again by c,
- then reciprocate,
- then multiply by $bc-ad$,
- and finally add $\frac{a}{c}$,

and the result is that x goes to $\frac{a}{c} + \frac{bc-ad}{c(cx+d)} = \frac{ax+b}{cx+d}$. When $c=0$, the linear fractional function is simply $\frac{ax+b}{cx+d} = \frac{a}{d}x + \frac{b}{d}$, and this can be composed from x by multiplying by a/d and then adding b/d.

Thus, *any linear fractional function is composed from the functions that reciprocate, multiply by k, and add l*, and hence (by the constructions in the previous section) *any linear fractional function on the number line is realized by a sequence of projections of the line.*

We now wish to prove the converse: *Any sequence of projections of the number line realizes a linear fractional function.* From the previous section, we know this is true for projection of a line onto a parallel line, so it suffices to find the function realized by projection of a line onto an intersecting line. We first take the case in which the lines are perpendicular (Figure 5.17). This case generalizes that of Figure 5.15, by allowing projection from an arbitrary point (a,b).

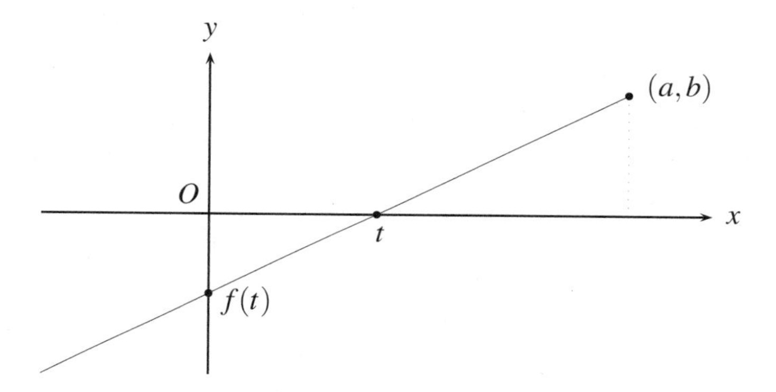

Figure 5.17: Projecting a line onto a perpendicular line

To find where the point t on the x-axis goes on the y-axis, we consider the slope of the line through t and (a,b). Between these points, the rise is b and the run is $a-t$, so the slope is $\frac{b}{a-t}$. Between t and the point $f(t)$ on the y-axis, the run is t and the rise is $-f(t)$; hence,

$$f(t) = \frac{bt}{t-a}, \quad \text{which is a linear fractional function.}$$

For the general case of intersecting lines, we take one line to be the x-axis again, and the other to be the line $y = cx$. Again we project the point t on the x-axis from (a,b) to the other line, and to find where t goes, we first find the equation of the line through t and (a,b). Equating the slope from t to (a,b) with the slope between an arbitrary point (x,y) on the line and (a,b), we find the equation

$$\frac{b}{a-t} = \frac{b-y}{a-x}.$$

This line meets the line $y = cx$ where

$$\frac{b}{a-t} = \frac{b-cx}{a-x},$$

and hence where

$$x = \frac{bt}{ct - ac + b}, \quad \text{which is also a linear fractional function of } t.$$

Thus, any single projection of a line can be represented by a linear fractional function of distance along the line. It is easy to check (Exercise 5.6.2) that the result of composing linear fractional functions is linear fractional. Hence, any finite sequence of projections is represented by a linear fractional function. □

Dividing by zero

You remember from high-school algebra that division by zero is not a valid operation, because it leads from true equations, such as $3 \times 0 = 2 \times 0$, to false ones, such as $3 = 2$. Nevertheless, *in carefully controlled situations*, it is permissible, and even enlightening, to divide by zero. One such situation is in projective mappings of the projective line.

The linear fractional functions $f(x) = \frac{ax+b}{cx+d}$ we have used to describe projective mappings of lines are actually defective if the variable x runs only through the set $\mathbb{R}$ of real numbers. For example, the function $f(x) = 1/x$ we used to map points of the line $\mathscr{L}_1$ onto points of the line $\mathscr{L}_2$ as shown in Figure 5.15 does not in fact map *all* points. It cannot send the point $x = 0$ anywhere, because $1/0$ is undefined; nor can it send any point to $y = 0$, because $0 \neq 1/x$ for any real x. This defect is neatly fixed by extending the function $f(x) = 1/x$ to a new object $x = \infty$, and declaring that $1/\infty = 0$ and $1/0 = \infty$. The new object ∞ is none other than the *point at infinity* of the line $\mathscr{L}_1$, which is supposed to map to the point 0 on $\mathscr{L}_2$. Likewise, if $1/0 = \infty$, the point 0 on $\mathscr{L}_1$ is sent to the point ∞ on $\mathscr{L}_2$, as it should be.

Thus, *the function $f(x) = 1/x$ works properly, not on the real line $\mathbb{R}$, but on the real projective line $\mathbb{R} \cup \{\infty\}$*—a line together with a point at infinity. The rules $1/\infty = 0$ and $1/0 = \infty$ simply reflect this fact.

It is much the same with any linear fractional function $f(x) = \frac{ax+b}{cx+d}$. The denominator of the fraction is 0 when $x = -d/c$, and the correct value of the function in this case is ∞. Conversely, no real value of x gives $f(x)$ the value a/c, but $x = \infty$ does. For this reason, *any function $f(x) = \frac{ax+b}{cx+d}$ with $ad - bc \neq 0$ maps the real projective line $\mathbb{R} \cup \{\infty\}$ onto itself.* The map is also one-to-one, as may seen in the exercises below.

The real projective line $\mathbb{RP}^1$

We can now give an algebraic definition of the object we called the "real projective line" in Section 5.3. It is the set $\mathbb{R} \cup \{\infty\}$ *together with* all the linear fractional functions mapping $\mathbb{R} \cup \{\infty\}$ onto itself. We call this set, with these functions on it, the *real projective line* $\mathbb{RP}^1$.

The set $\mathbb{R} \cup \{\infty\}$ certainly has the points we require for a projective line; the functions are to give $\mathbb{R} \cup \{\infty\}$ the "elasticity" of a line that undergoes projection. The ordinary line $\mathbb{R}$ is not very "elastic" in this sense. Once we have decided which point is 0 and which point is 1, the numerical value of every point on $\mathbb{R}$ is uniquely determined. In contrast, the position of a point on $\mathbb{RP}^1$ is *not* determined by the positions of 0 and 1 alone.

For example, there is a projection that sends 0 to 0, 1 to 1, but 2 to 3. Nevertheless, there is a constraint on the "elasticity" of $\mathbb{RP}^1$. If 0 goes to 0, 1 goes to 1, and 2 goes to 3, say, then the destination of every *other* point x is uniquely determined. In the next two sections, we will see why.

Exercises

The formula $\frac{ax+b}{cx+d} = \frac{a}{c} + \frac{bc-ad}{c(cx+d)}$ gives an inkling why the condition $ad - bc \neq 0$ is part of the definition of a linear fractional function: If $ad - bc = 0$, then $\frac{ax+b}{cx+d} = \frac{a}{c}$ is a constant function, and hence it maps the whole line onto one point.

If we want to map the line onto another line, it is therefore necessary to have $ad - bc \neq 0$. It is also sufficient, because we can solve the equation $y = \frac{ax+b}{cx+d}$ for x in that case.

5.6.1 Solve the equation $y = \frac{ax+b}{cx+d}$ for x, and note where your solution assumes $ad - bc \neq 0$.

5.6.2 If $f_1(x) = \frac{a_1x+b_1}{c_1x+d_1}$ and $f_2(x) = \frac{a_2x+b_2}{c_2x+d_2}$, compute $f_1(f_2(x))$, and verify that it is of the form $\frac{Ax+B}{Cx+D}$.

5.6.3 Verify also that $\begin{pmatrix} A & B \\ C & D \end{pmatrix} = \begin{pmatrix} a_1 & b_1 \\ c_1 & d_1 \end{pmatrix} \begin{pmatrix} a_2 & b_2 \\ c_2 & d_2 \end{pmatrix}$.

Thus, linear fractional functions behave like 2×2 matrices. Moreover, the condition $ad - bc \neq 0$ corresponds to *having nonzero determinant*, which explains why this is the condition for an inverse function to exist.

5.6.4 It also guarantees that if $a_1d_1 - b_1c_1 \neq 0$ for $f_1(x)$ and $a_2d_2 - b_2c_2 \neq 0$ for $f_2(x)$ in Exercise 5.5.2, then $AD - BC \neq 0$ for $f_1(f_2(x))$. Why?

5.7 The cross-ratio

> You might say it was a triumph of algebra to invent this quantity that turns out to be so valuable and could not be imagined geometrically. Or if you are a geometer at heart, you may say it is an invention of the devil and hate it all your life.
>
> Robin Hartshorne, *Geometry: Euclid and Beyond*, p. 341.

It is visually obvious that projection can change lengths and even the ratio of lengths, because equal lengths often appear unequal under projection. And yet we can recognize that Figure 5.1 is a picture of equal tiles, even though they are unequal in size and shape. Some clue to their equality must be preserved, but what? It cannot be length; it cannot be a ratio of lengths; but, surprisingly, it can be a *ratio of ratios*, called the *cross-ratio*.

The cross-ratio is a quantity associated with four points on a line. If the four points have coordinates p, q, r, and s, then their cross-ratio is the function of the ordered 4-tuple (p,q,r,s) defined by

$$\frac{(r-p)/(s-p)}{(r-q)/(s-q)}, \quad \text{which can also be written as} \quad \frac{(r-p)(s-q)}{(r-q)(s-p)}.$$

The cross-ratio is preserved by projection. To show this, it suffices to show that it is preserved by the three generating transformations from which we composed all linear fractional maps in the previous section:

1. The map sending x to $x+l$.

   Here the numbers p,q,r,s are replaced by $p+l, q+l, r+l, s+l$, respectively. This does not change the cross-ratio because the l terms cancel by subtraction.

2. The map sending x to kx.

   Here the numbers p,q,r,s are replaced by kp, kq, kr, ks, respectively. This does not change the cross-ratio because the k terms cancel by division.

3. The map sending x to $1/x$.

   Here the numbers p,q,r,s are replaced by $\frac{1}{p}, \frac{1}{q}, \frac{1}{r}, \frac{1}{s}$, respectively, so the cross-ratio

   $$\frac{(r-p)(s-q)}{(r-q)(s-p)}$$

is replaced by

$$\frac{\left(\frac{1}{r}-\frac{1}{p}\right)\left(\frac{1}{s}-\frac{1}{q}\right)}{\left(\frac{1}{r}-\frac{1}{q}\right)\left(\frac{1}{s}-\frac{1}{p}\right)}=\frac{\frac{p-r}{pr}\cdot\frac{q-s}{qs}}{\frac{q-r}{qr}\cdot\frac{p-s}{ps}} \quad \text{taking common denominators,}$$

$$=\frac{(p-r)(q-s)}{(q-r)(p-s)} \quad \text{multiplying through by } pqrs,$$

$$=\frac{(r-p)(s-q)}{(r-q)(s-p)} \quad \text{changing the sign in all factors,}$$

and thus, the cross-ratio is unchanged in this case also. □

Is the cross-ratio visible?

If we take the four equally spaced points $p=0$, $q=1$, $r=2$, and $s=3$ on the line, then their cross-ratio is

$$\frac{(r-p)(s-q)}{(r-q)(s-p)}=\frac{2\times 2}{1\times 3}=\frac{4}{3}.$$

It follows that any projective image of these points also has cross-ratio $4/3$. Do four points on a line *look* equally spaced if their cross-ratio is 4/3? Test your eye on the quadruples of points in Figure 5.18, and then do Exercise 5.7.2 to find the correct answer.

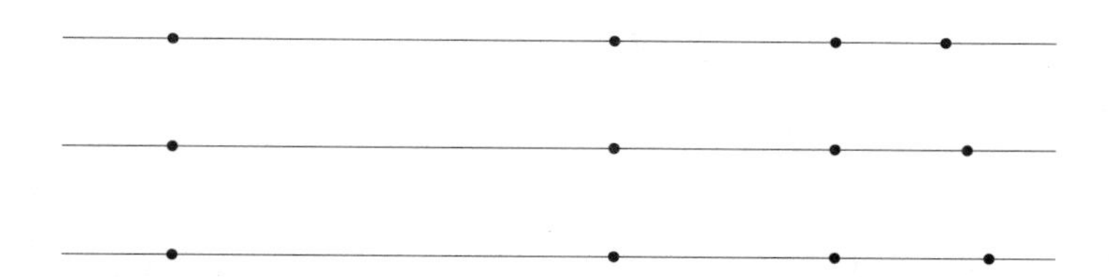

Figure 5.18: Which is a projective image of equally spaced points?

Exercises

We will see in the next section that any three points on $\mathbb{RP}^1$ can be projected to any three points. Hence, there cannot be an invariant involving just three points.

However, the invariance of the cross-ratio tells us that, once the images of three points are known, the whole projection map is known (compare with the "three-point determination" of isometries of the plane in Section 3.7).

5.7.1 Show that there is only one point s that has a given cross-ratio with given points p, q, and r.

In particular, if we have the points $p=0$, $q=2$, $r=3$ (which we do in the three quadruples in Figure 5.18), there is exactly one s that gives the cross-ratio 4/3 required for "equally spaced" points.

5.7.2 Find the value of s that gives the cross-ratio 4/3, and hence find the "equally spaced" quadruple in Figure 5.18.

Before the discovery of perspective, artists sometimes attempted to draw a tiled floor by making the width of each row of tiles a constant fraction e of the one before.

5.7.3 Show that this method is not correct by computing the cross-ratio of four points separated by the distances 1, e, and e^2.

5.8 What is special about the cross-ratio?

In the remainder of this book, we use the abbreviation

$$[p,q;r,s] = \frac{(r-p)(s-q)}{(r-q)(s-p)}$$

for the cross-ratio of the four points p,q,r,s, taken in that order.

We have shown that the cross-ratio is an invariant of linear fractional transformations, but it is obviously not the only one. Examples of other invariants are $(\text{cross-ratio})^2$ and cross-ratio $+1$. The cross-ratio is special because it is the *defining invariant* of linear fractional transformations. That is, *the linear fractional transformations are precisely the transformations of* $\mathbb{RP}^1$ *that preserve the cross-ratio.* (Thus, the cross-ratio defines linear fractional transformations the way that length defines isometries.)

We prove this fact among several others about linear fractional transformations and the cross-ratio.

Fourth point determination. *Given any three points* $p,q,r \in \mathbb{RP}^1$, *any other point* $x \in \mathbb{RP}^1$ *is uniquely determined by its cross-ratio* $[p,q;r,x] = y$ *with* p,q,r.

This statement holds because we can solve the equation

$$y = \frac{(r-p)(x-q)}{(r-q)(x-p)}$$

uniquely for x in terms of p,q,r, and y. □

Existence of three-point maps. *Given three points $p, q, r \in \mathbb{RP}^1$ and three points $p', q', r' \in \mathbb{RP}^1$, there is a linear fractional transformation f sending p, q, r to p', q', r', respectively.*

This statement holds because there is a projection sending any three points p, q, r to any three points p', q', r', and any projection is linear fractional by Section 5.6. The way to project is shown in Figure 5.19.

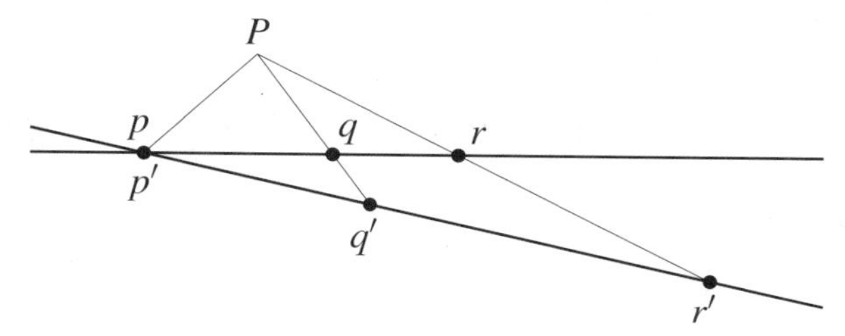

Figure 5.19: Projecting three points to three points

Without loss of generality, we can place the two copies of $\mathbb{RP}^1$ so that $p = p'$. Then the required projection is from the point P where the lines qq' and rr' meet. □

Uniqueness of three-point maps. *Exactly one linear fractional function sends three points p, q, r to three points p', q', r', respectively.*

A linear fractional f sending p, q, r to p', q', r', respectively, must send any $x \neq p, q, r$ to x' satisfying $[p, q; r, x] = [p', q'; r', x']$, because f preserves the cross-ratio by Section 5.7. But x' is unique by fourth point determination, so there is exactly one such function f. □

Characterization of linear fractional maps. *These are precisely the maps of $\mathbb{RP}^1$ that preserve the cross-ratio.*

By Section 5.7, any linear fractional map f preserves the cross-ratio. That is, $[f(p, f(q); f(r), f(s)] = [p, q; r, s]$ for any four points p, q, r, s.

Conversely, suppose that f is a map of $\mathbb{RP}^1$ with

$$[f(p), f(q); f(r), f(s)] = [p, q; r, s] \quad \text{for any four points } p, q, r, s.$$

By the existence of three-point maps, we can find a linear fractional g that agrees with f on p, q, r. But then, because f preserves the cross-ratio, g agrees with f on s also, by unique fourth point determination.

Thus, g agrees with f everywhere, so f is a linear fractional map. □

The existence of three-point maps says that any three points on $\mathbb{RP}^1$ can be sent to any three points by a linear fractional transformation. Thus, any invariant of triples of points must have the *same value* for any triple, and so it is trivial. A nontrivial invariant must involve at least four points, and the cross-ratio is an example. It is in fact the *fundamental* example, in the following sense.

The fundamental invariant. *Any invariant of four points is a function of the cross-ratio.*

To see why, suppose $I(p,q,r,s)$ is a function, defined on quadruples of distinct points, that is invariant under linear fractional transformations. Thus,

$$I(f(p),f(q),f(r),f(s)) = I(p,q,r,s) \quad \text{for any linear fractional } f.$$

In other words, I has the same value on all quadruples (p',q',r',s') that result from (p,q,r,s) by a linear fractional transformation. But more is true: I *has the same value on all quadruples* (p',q',r',s') *with the same cross-ratio as* (p,q,r,s), because such a quadruple (p',q',r',s') results from (p,q,r,s) by a linear fractional transformation. This follows from the existence and uniqueness of three-point maps:

- by existence, we can send p,q,r to p',q',r', respectively, by a linear fractional transformation f, and
- by uniqueness, f also sends s to s', the unique point that makes $[p,q;r,s] = [p',q';r',s']$.

Because I has the same value on all quadruples with the same cross-ratio, it is meaningful to view I as a function J *of* the cross-ratio, defined by

$$J([p,q;r,s]) = I(p,q,r,s). \qquad \square$$

Exercises

The following exercises illustrate the result above about invariant functions of quadruples. They show that the invariants obtained by *permuting the variables* in the cross-ratio $y = [p,q;r,s]$ are simple functions of y, such as $1/y$ and $y-1$.

5.8.1 If $[p,q;r,s] = y$, show that $[p,q;s,r] = 1/y$.

5.8.2 If $[p,q;r,s] = y$, show that $[q,p;r,s] = 1/y$.

5.8.3 Prove that $[p,q;r,s]+[p,r;q,s]=1$, so it follows that if $[p,q;r,s]=y$, then $[p,r;q,s]=1-y$.

The transformations $y\mapsto 1/y$ and $y\mapsto 1-y$ obtained in this way generate all transformations of the cross-ratio obtained by permuting its variables. There are six such transformations (even though there are 24 permutations of four variables).

5.8.4 Show that the functions of y obtained by combining $1/y$ and $1-y$ in all ways are

$$y,\quad \frac{1}{y},\quad 1-y,\quad 1-\frac{1}{y},\quad \frac{1}{1-y},\quad \frac{y}{y-1}.$$

5.8.5 Explain why any permutation of four variables may be obtained by exchanges: either of the first two, the middle two, or the last two variables.

5.8.6 Deduce from Exercises 5.8.1–5.8.5 that the invariants obtained from the cross-ratio y by permuting its variables are precisely the six listed in Exercise 5.8.4.

The six linear fractional functions of y obtained in Exercise 5.8.4 constitute what is sometimes called the *cross-ratio group*. It is an example of a concept we will study in Chapter 7: the concept of a *group of transformations*. Unlike most of the groups studied there, this group is finite.

5.9 Discussion

The plane $\mathbb{RP}^2$ studied in this chapter is the most important projective plane, but it is far from being the only one. Many other projective planes can be constructed by imitating the construction of $\mathbb{RP}^2$, which is based on ordered triples (x,y,z) and linear equations $ax+by+cz=0$. It is not essential for x,y,z to be real numbers. As noted earlier, they could be complex numbers, but more generally they could be elements of any *field*. A field is any set with $+$ and $\times$ operations satisfying the nine field axioms listed in Section 4.8.

If $\mathbb{F}$ is any field, we can consider the space $\mathbb{F}^3$ of ordered triples (x,y,z) with $x,y,z\in\mathbb{F}$. Then the projective plane $\mathbb{FP}^2$ has

- "points," each of which is a set of triples (kx,ky,kz), where $x,y,z\in\mathbb{F}$ are fixed and k runs through the elements of $\mathbb{F}$,
- "lines," each of which consists of the "points" satisfying an equation of the form $ax+by+cz=0$ for some fixed $a,b,c\in\mathbb{F}$.

The projective plane axioms can be checked for $\mathbb{FP}^2$ just as they were for $\mathbb{RP}^2$. The same calculations apply, because the field axioms ensure that the same algebraic operations work in $\mathbb{F}$ (solving equations, for example). This gives a great variety of planes $\mathbb{FP}^2$, because there are a great variety of fields $\mathbb{F}$.

Perhaps the most familiar field, after $\mathbb{R}$ and $\mathbb{C}$, is the set $\mathbb{Q}$ of rational numbers. $\mathbb{QP}^2$ is not unlike $\mathbb{RP}^2$, except that all of its points have rational coordinates, and all of its lines are full of gaps, because they contain only rational points.

More surprising examples arise from taking $\mathbb{F}$ to be a *finite* field, of which there is one with p^n elements for each power p^n of each prime p. The simplest example is the field $\mathbb{F}_2$, whose members are the elements 0 and 1, with the following addition and multiplication tables.

$+$	0	1
0	0	1
1	1	0

$\times$	0	1
0	0	0
1	0	1

The projective plane $\mathbb{F}_2\mathbb{P}^2$ has seven points, corresponding to the seven nonzero points in $\mathbb{F}_2^3$:

$$(1,0,0),\ (0,1,0),\ (0,0,1),\ (0,1,1),\ (1,0,1),\ (1,1,0),\ (1,1,1).$$

These points are arranged in threes along the seven lines in Figure 5.20, one of which is drawn as a circle so as to connect its three points.

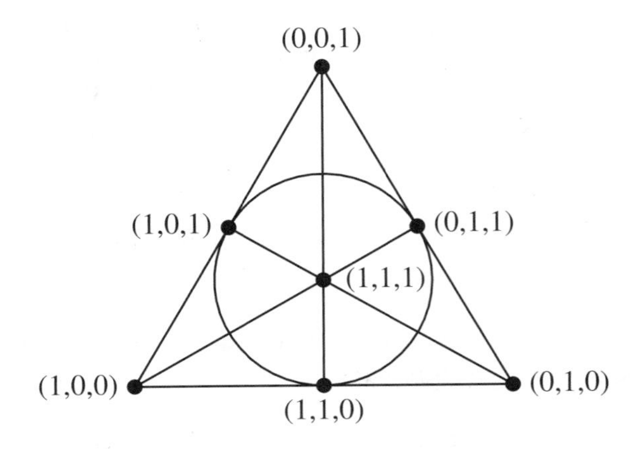

Figure 5.20: The smallest projective plane

Notice that the lines satisfy the linear equations

$$x=0, \quad y=0, \quad z=0,$$
$$x+y=0, \quad y+z=0, \quad z+x=0,$$
$$x+y+z=0.$$

For example, the points on the circle satisfy $x+y+z=0$. (Of course, the coordinates have nothing to do with position in the plane of the diagram. The figure is mainly symbolic, while attempting to show "points" collected into "lines.")

This structure is called the *Fano plane*, and it is the smallest projective plane. Despite being small, it is well-behaved, because its "lines" satisfy linear equations, just as lines do in the traditional geometric world. Thanks to finite fields, linear algebra works well in many finite structures. It has led to the wholesale development of finite geometries, many of which have applications in the mathematics of information and communication.

However, the three axioms for a finite projective plane do not ensure that the plane is of the form $\mathbb{FP}^2$, with coordinates for points and linear equations for lines. They can be satisfied by bizarre "nonlinear" structures, as we will see in the next chapter. A fourth axiom is needed to engender a field $\mathbb{F}$ of coordinates, and the axiom is none other than the theorem of Pappus that we met briefly in Chapters 1 and 4. This state of affairs will be explained in Chapter 6.

The invariance of the cross-ratio

The invariance of the cross-ratio was discovered by Pappus around 300 CE and rediscovered by Desargues around 1640. It appears (not very clearly) as Proposition 129 in Book VII of Pappus' *Mathematical Collection* and again in *Manière universelle de Mr Desargues* in 1648. The latter is a pamphlet on perspective by written by Abraham Bosse, a disciple of Desargues. It also contains the first published statement of the Desargues theorem mentioned in Chapters 1 and 4. Because of this, and the fact that he wrote the first book on projective geometry, Desargues is considered to be the founder of the subject. Nevertheless, projective geometry was little known until the 19th century, when geometry expanded in all directions. In the more general 19th century geometry (which often included use of complex numbers), the cross-ratio continued to be a central concept.

One of the reasons we now consider the appropriate generalization of classical projective geometry to be projective geometry with coordinates in a field is that *the cross-ratio continues to make sense in this setting.*

Linear fractional transformations and the cross-ratio make sense when $\mathbb{R}$ is replaced by any field $\mathbb{F}$. The transformations $x \mapsto x + l$ and $x \mapsto kx$ make sense on $\mathbb{F}$, and $x \mapsto 1/x$ makes sense on $\mathbb{F} \cup \{\infty\}$ if we set $1/0 = \infty$ and $1/\infty = 0$. Then the transformations

$$x \mapsto \frac{ax+b}{cx+d}, \quad \text{where } a,b,c,d \in \mathbb{F} \text{ and } ad - bc \neq 0,$$

make sense on the "$\mathbb{F}$ *projective line*" $\mathbb{FP}^1 = \mathbb{F} \cup \{\infty\}$. The cross-ratio is invariant by the same calculation as in Section 5.7, thanks to the field axioms, because the usual calculations with fractions are valid in a field.

6

Projective planes

PREVIEW

In this chapter, geometry fights back against the forces of arithmetization. We show that coordinates need not be brought into geometry from outside—they can be defined by purely geometric means. Moreover, the geometry required to define coordinates and their arithmetic is *simpler* than Euclid's geometry. It is the *projective geometry* introduced in the previous chapter, but we have to build it from scratch using properties of straight lines alone.

We started this project in Section 5.3 by stating the three axioms for a projective plane. However, these axioms are satisfied by many structures, some of which have no reasonable system of coordinates. To build coordinates, we need at least one additional axiom, but for convenience we take two: the *Pappus* and *Desargues* properties that were proved with the help of coordinates in Chapter 4.

Here we proceed in the direction opposite to Chapter 4: Take Pappus and Desargues as axioms, and use them to define coordinates. The coordinates are points on a projective line, and we add and multiply them by constructions like those in Chapter 1. But instead of using parallel lines as we did there, we call lines "parallel" if they meet on a designated line called the "horizon" or the "line at infinity."

The main problem is to prove that our addition and multiplication operations satisfy the field axioms. This is where the theorems of Pappus and Desargues are crucial. Pappus is needed to prove the *commutative law* of multiplication, $ab = ba$, whereas Desargues is needed to prove the *associative law*, $a(bc) = (ab)c$.

6.1 Pappus and Desargues revisited

The theorems of Pappus and Desargues stated in Chapters 1 and 4 had a similar form: If two particular pairs of lines are parallel, then a third pair is parallel. Because parallel lines meet on the horizon, the Pappus and Desargues theorems also say that if two particular pairs of lines meet on the horizon, then so does a third pair. And because the horizon is not different from any other line, these theorems are really about three pairs of lines having their intersections on the *same* line.

In this projective setting, the Pappus theorem takes the form shown in Figure 6.1. The six vertices of the hexagon are shown as dots, and the opposite sides are shown as a black pair, a gray pair, and a dotted pair. The line on which each of the three pairs meet is labeled $\mathscr{L}$, and we have oriented the figure so that $\mathscr{L}$ is horizontal (but this is not at all necessary).

Projective Pappus theorem. *Six points, lying alternately on two straight lines, form a hexagon whose three pairs of opposite sides meet on a line.*

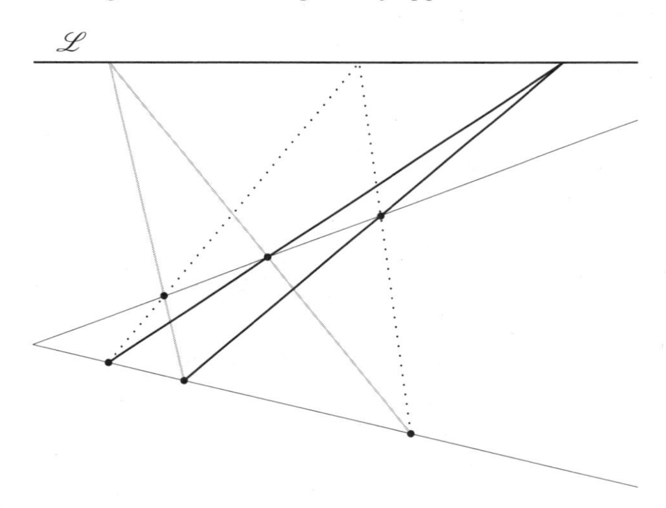

Figure 6.1: The projective Pappus configuration

This statement of the Pappus theorem is called *projective* because it involves only the concepts of points, lines, and meetings between them. Meetings between geometric objects are called *incidences*, and, for this reason, the Pappus theorem is also called an *incidence theorem*. The three axioms of a projective plane, given in Section 5.3, are the simplest examples of incidence theorems.

The projective Desargues theorem is another incidence theorem. It concerns the pairs of corresponding sides of two triangles, shown in solid gray in Figure 6.2. The triangles are in *perspective from a point P*, which means that each pair of corresponding vertices lies on a line through P. The three corresponding pairs of sides are again shown as black, gray, and dotted, and each pair meets on a line labeled $\mathscr{L}$.

Projective Desargues theorem. *If two triangles are in perspective from a point, then their pairs of corresponding sides meet on a line.*

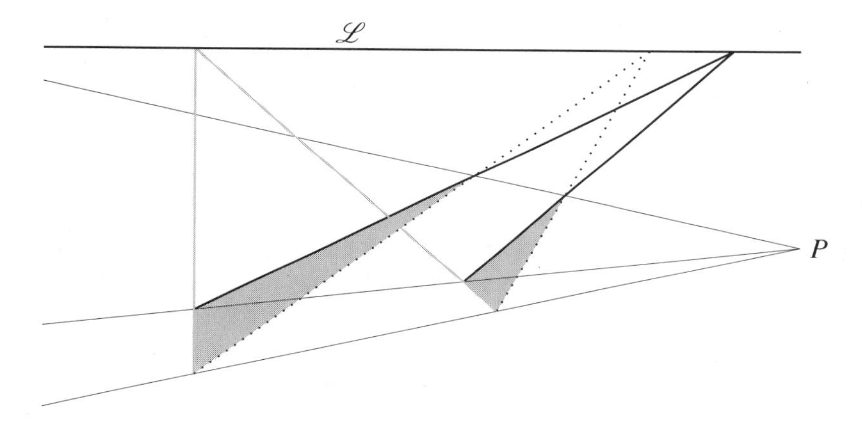

Figure 6.2: The projective Desargues configuration

An important special case of the Desargues theorem has the center of projection P on the line $\mathscr{L}$ where the corresponding sides of the triangles meet. This special case is called the *little Desargues theorem*, and it is shown in Figure 6.3.

Little Desargues theorem. *If two triangles are in perspective from a point P, and if two pairs of corresponding sides meet on a line* $\mathscr{L}$ *through P, then the third pair of corresponding sides also meets on* $\mathscr{L}$.

Because the projective Pappus and Desargues theorems involve only incidence concepts, one would like proofs of them that involve only the three axioms for a projective plane given in Section 5.3. Unfortunately, this is not possible, because there are examples of projective planes *not* satisfying the Pappus and Desargues theorems. What we can do, however, is take the Pappus and Desargues theorems as *new axioms*. Together with the original three axioms for projective planes, these two new axioms apply to a broad class of projective planes called *Pappian planes*.

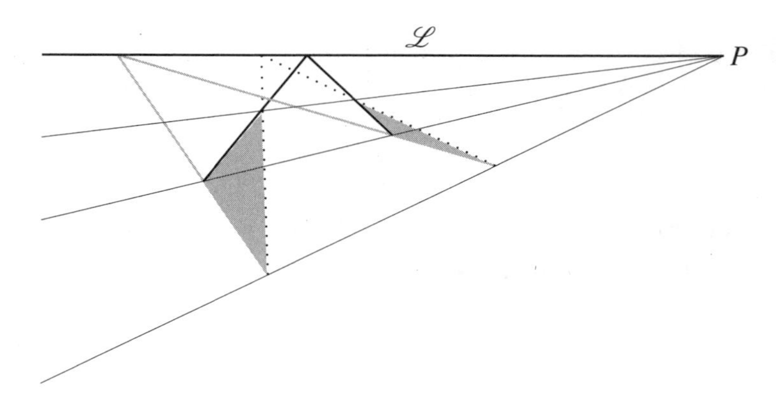

Figure 6.3: The little Desargues configuration

The Pappian planes include $\mathbb{RP}^2$ and many other planes, but not all. They turn out to be the planes with *coordinates* satisfying the same *laws of algebra* as the real numbers—the field axioms. The object of this chapter is to show how coordinates arise when the Pappus and Desargues theorems hold, and why they satisfy the field axioms. In doing so, we will see that projective geometry is *simpler* than algebra in a certain sense, because we use only five geometric axioms to derive the nine field axioms.

Exercises

In some projective planes, the Desargues theorem is false. Here is one example, which is called the *Moulton plane*. Its "points" are ordinary points of $\mathbb{R}^2$, together with a point at infinity for each family of parallel "lines." However, the "lines" of the Moulton plane are not all ordinary lines. They include the ordinary lines of negative, horizontal, or vertical slope, but each other "line" is a *broken line* consisting of a half line of slope $k > 0$ below the x-axis, joined to a half line of slope $k/2$ above the x-axis. Figure 6.4 shows some of the "lines."

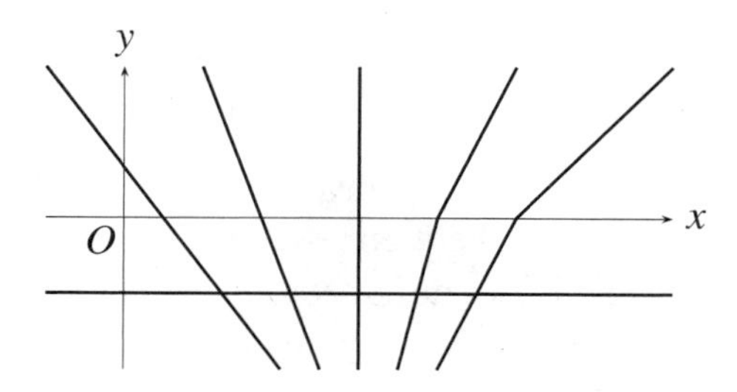

Figure 6.4: Lines of the Moulton plane

6.1.1 Find where the "line" from $(0,-1)$ to $(2,1/2)$ meets the x-axis.

6.1.2 Explain why any two "points" of the Moulton plane lie on a unique "line."

6.1.3 Explain why any two "lines" of the Moulton plane meet in a unique "point." (Parallel "lines" have a common "point at infinity" by definition, so do not worry about them.)

6.1.4 Give four "points," no three of which lie on the same "line."

6.1.5 Thus, the Moulton plane satisfies the three axioms of a projective plane. But it does not satisfy even the little Desargues theorem, as Figure 6.5 shows. Explain.

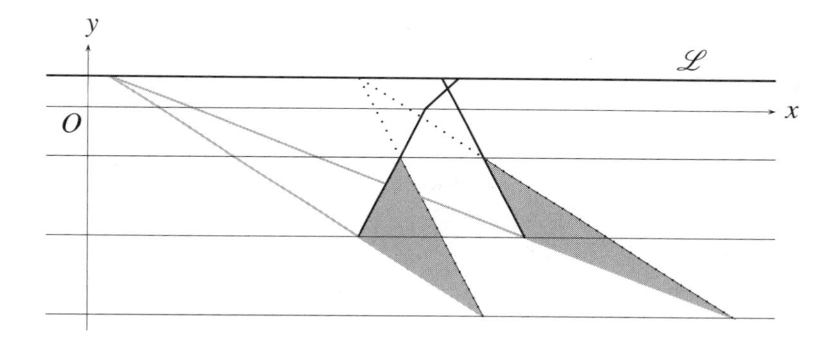

Figure 6.5: Failure of the little Desargues theorem in the Moulton plane

6.2 Coincidences

Two points A, B always lie on a line. But it is accidental, so to speak, if a third point C lies on the line through A and B. Such an accidental meeting is called a "coincidence" in everyday life, and this is a good name for it in projective geometry too: *co*incidence = two incidences together—in this case the incidence of A and B with a line, and the incidence of C with the same line.

The theorems of Pappus and Desargues state that certain coincidences occur. In fact, they are coincidences of the type just described, in which two points lie on a line and a third point lies on the same line. The perspective picture of the tiled floor also involves certain coincidences, as becomes clear when we look again at the first few steps in its construction (Figure 6.6).

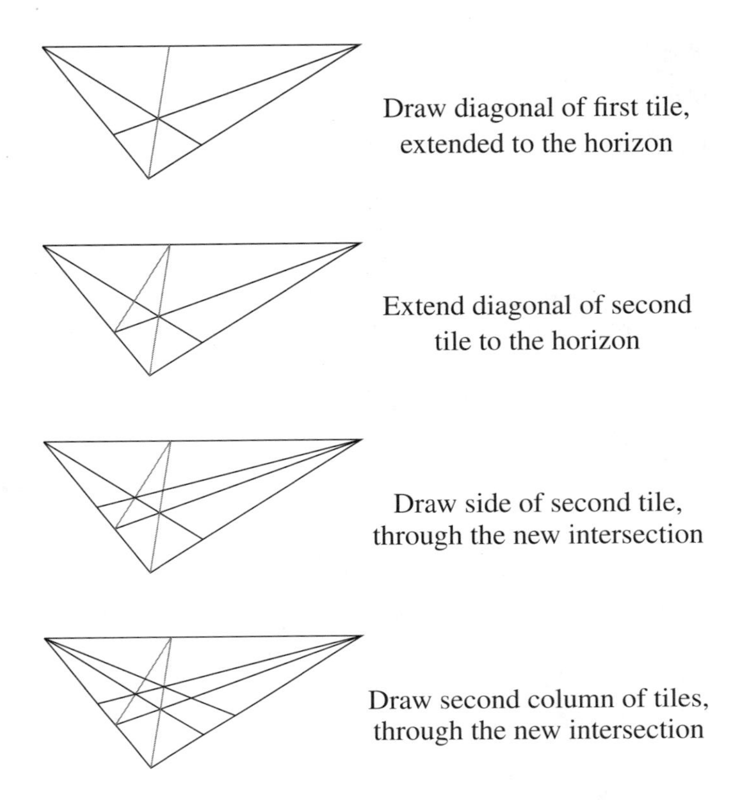

Figure 6.6: Constructing the tiled floor

At this step, a coincidence occurs. Three of the points we have constructed lie on a straight line, which is shown dashed in Figure 6.7.

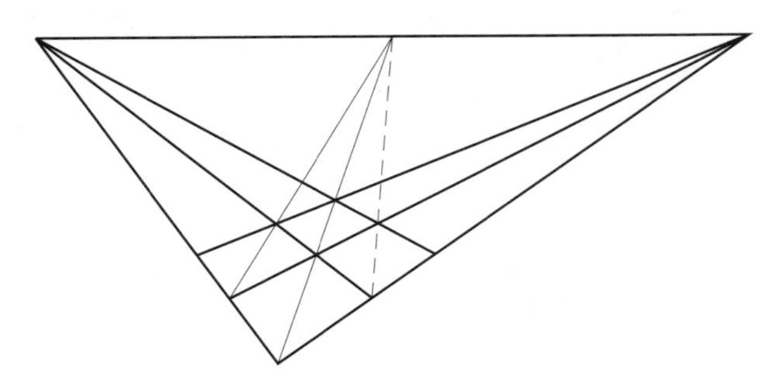

Figure 6.7: A coincidence in the tiled floor

This coincidence can be traced to a special case of the little Desargues theorem, which involves the two shaded triangles shown in Figure 6.8.

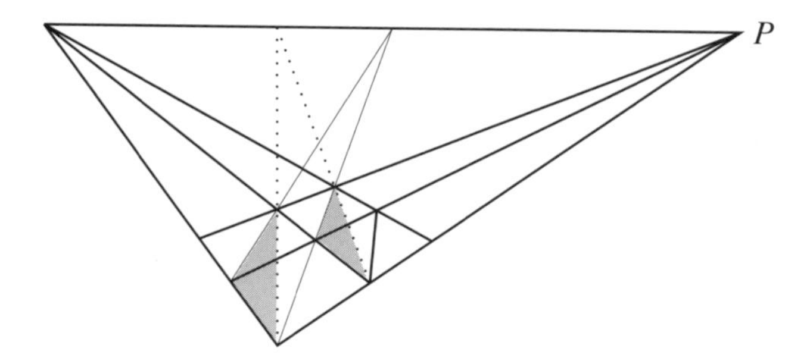

Figure 6.8: A little Desargues configuration in the tiled floor

This case of little Desargues says that the two dotted lines (diagonals of "double tiles") meet on the horizon. These lines give us a second little Desargues configuration, shown in Figure 6.9, from which we conclude that the dashed diagonals also meet on the horizon, as required to explain the coincidence in Figure 6.7.

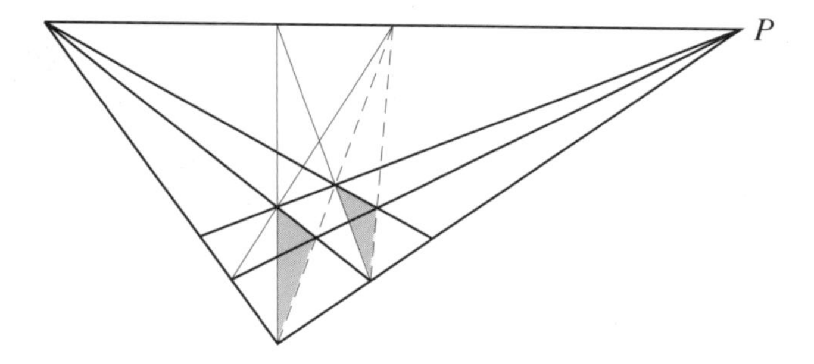

Figure 6.9: A second little Desargues configuration

Exercises

The occurrence of the little Desargues configuration in the tiled floor may be easier to see if we draw the lines meeting on the horizon as actual parallels. The little Desargues theorem itself is easier to state in terms of actual parallels (Figure 6.10).

6.2.1 Formulate an appropriate statement of the little Desargues theorem when one has parallels instead of lines meeting on $\mathscr{L}$.

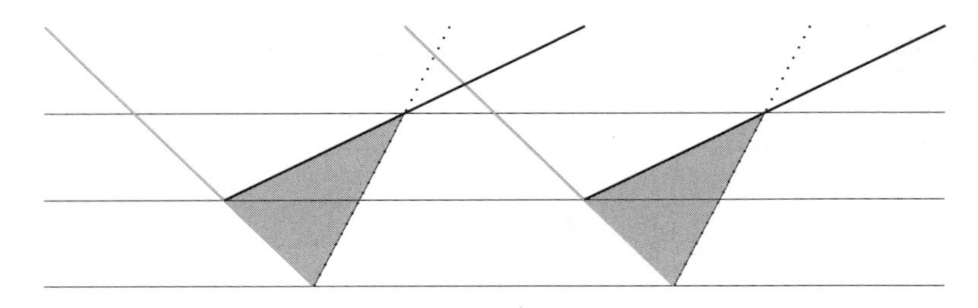

Figure 6.10: The parallel little Desargues configuration

6.2.2 Now redraw Figures 6.7, 6.8, and 6.9 so that the lines meeting on $\mathcal{L}$ are shown as actual parallels.

6.2.3 What is the nature of the "coincidence" in Figure 6.7 now?

6.2.4 Find occurrences of the little Desargues configuration in your diagrams. Hence, explain why the "coincidence" in Exercise 6.2.3 follows from your statement of the little Desargues theorem in Exercise 6.2.1.

The theorem that proves the coincidence in the drawing of the tiled floor is actually a special case of the little Desargues theorem: the case in which a vertex of one triangle lies on a side of the other. Thus, it is not clear that the coincidence is false in the Moulton plane, where we know only that the general little Desargues theorem is false by the exercises in Section 6.1.

6.2.5 By placing an x-axis in a suitable position on Figure 6.11, show that the tiled floor coincidence fails in the Moulton plane.

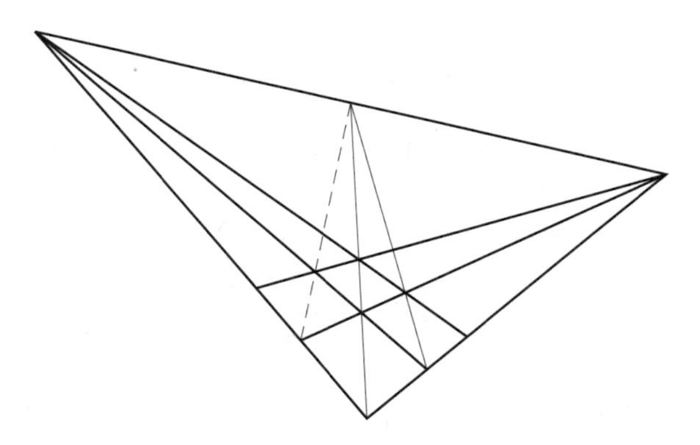

Figure 6.11: A coincidence that fails in the Moulton plane

6.3 Variations on the Desargues theorem

In Section 6.1, we stated the Desargues theorem in the form: *If two triangles are in perspective from a point, then their three pairs of corresponding sides meet on a line.* The Desargues theorem is a very flexible theorem, which appears in many forms, and two that we need later are the following. (We need these theorems only as consequences of the Desargues theorem, but they are actually equivalent to it.)

Converse Desargues theorem. *If corresponding sides of two triangles meet on a line, then the two triangles are in perspective from a point.*

To deduce this result from the Desargues theorem, let ABC and $A'B'C'$ be two triangles whose corresponding sides meet on the line $\mathscr{L}$. Let P be the intersection of AA' and BB', so we want to prove that P lies on CC' as well. Suppose that PC meets the line $B'C'$ at C'' (Figure 6.12 shows C'', hypothetically, unequal to C').

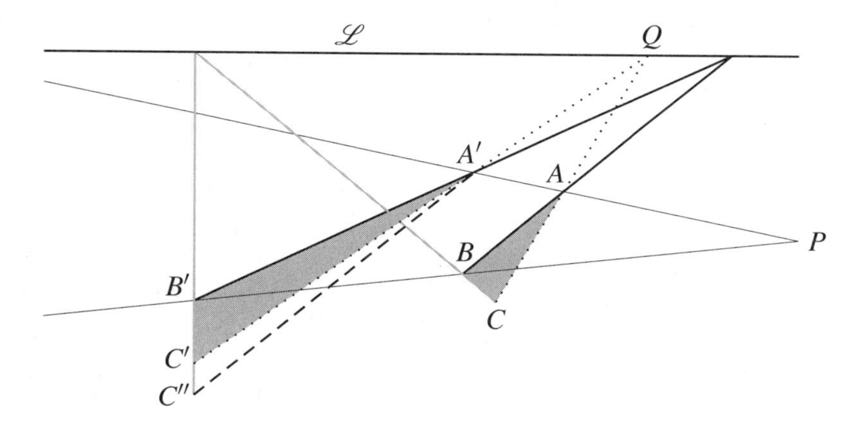

Figure 6.12: The converse Desargues theorem

Then the triangles ABC and $A'B'C''$ are in perspective from P and therefore, by the Desargues theorem, their corresponding sides meet on a line. We already know that AB meets $A'B'$ on $\mathscr{L}$, and that BC meets $B'C'$ on $\mathscr{L}$. Hence, AC meets $A'C''$ on $\mathscr{L}$, necessarily at the point Q where AC meets $\mathscr{L}$. It follows that QA' goes through C''. But we also know that QA' meets $B'C'$ at C'. Hence, $C'' = C'$.

Thus, C' is indeed on the line PC, so ABC and $A'B'C'$ are in perspective from P, as required. □

The second consequence of the Desargues theorem is called the "scissors theorem." I do not know how common this name is, but it is used on p. 69 of the book *Fundamentals of Mathematics II. Geometry*, edited by Behnke, Bachmann, Fladt, and Kunle. In any case, it is an apt name, as you will see from Figure 6.13.

Scissors theorem. *If $ABCD$ and $A'B'C'D'$ are quadrilaterals with vertices alternately on two lines, and if AB is parallel to $A'B'$, BC to $B'C'$, and AD to $A'D'$, then also CD is parallel to $C'D'$.*

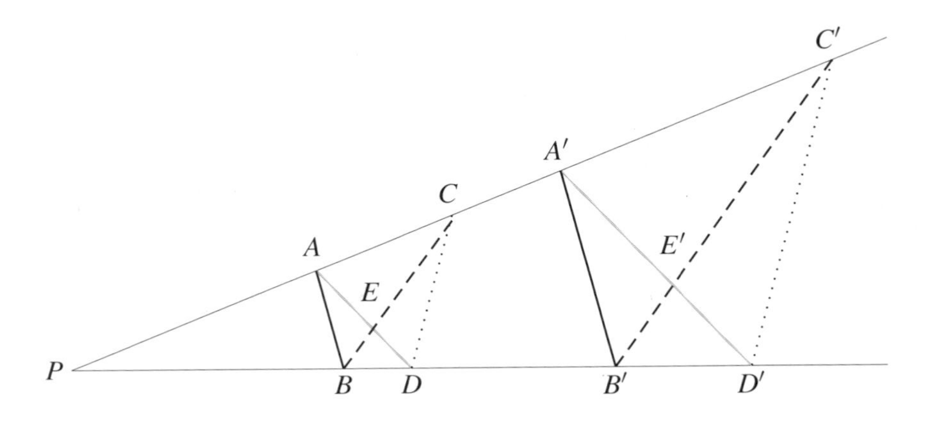

Figure 6.13: The scissors theorem

To prove this theorem, let E be the intersection of AD and BC and let E' be the intersection of $A'D'$ and $B'C'$, as shown in Figure 6.13. Then the triangles ABE and $A'B'E'$ have corresponding sides parallel. Hence, they are in perspective from the intersection P of AA' and BB', by the converse Desargues theorem.

But then the triangles CDE and $C'D'E'$ are also in perspective from P. Because their sides CE and $C'E'$, DE and $D'E'$, are parallel by assumption, it follows from the Desargues theorem that CD and $C'D'$ are also parallel, as required. □

The scissors theorem just proved says that if the black, gray, and dashed lines in Figure 6.13 are parallel, then so are the dotted lines. *What if the black, gray, and dotted lines are parallel: Are the dashed lines again parallel?* The answer is yes, and the proof is similar, but with a slightly different picture (Figure 6.14).

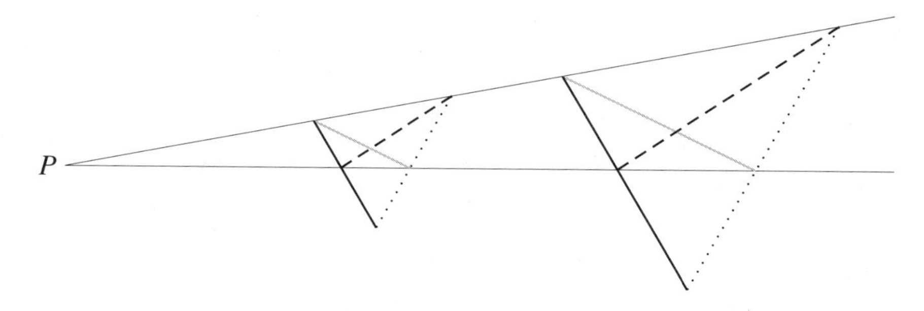

Figure 6.14: Second case of the scissors theorem

We have extended the black and dotted lines until they meet, forming triangles with their corresponding black, gray, and dotted sides parallel. Then it follows from the converse Desargues theorem that these triangles are in perspective from P. But then so are the triangles with dashed, black, and dotted sides. Hence their dashed sides are parallel by the Desargues theorem. □

Remark. In practice, the scissors theorem is often used in the following way. We have a pair of scissors $ABCD$ and another figure $D'A'B'C'F'$ with parallel pairs of black, gray, dashed, and dotted lines as shown in Figure 6.15. We want to prove that $D' = F'$ (so the ends of the gray and dotted lines *coincide*, and the second figure is also a pair of scissors).

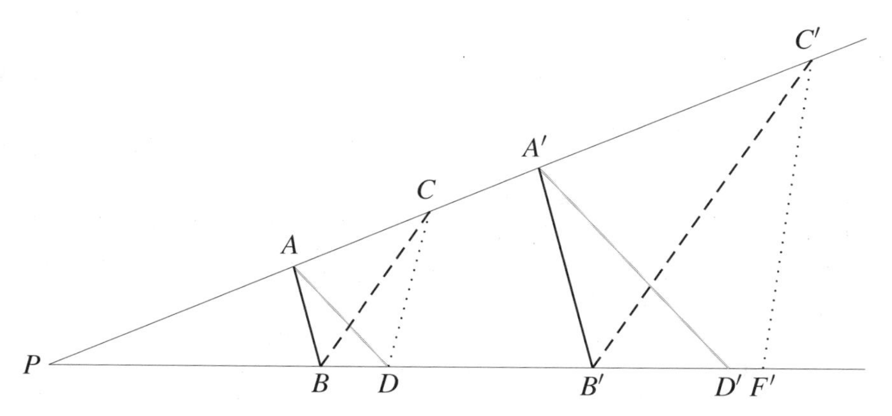

Figure 6.15: Applying the scissors theorem

This coincidence happens because the line $C'D'$ is parallel to CD by the scissors theorem, so $C'D'$ is the *same* line as $C'F'$, and hence $D' = F'$.

Exercises

Because the Desargues theorem implies its converse, another way to show that the Desargues theorem fails in the Moulton plane is to show that its converse fails. This plan is easily implemented with the help of Figure 6.16. (Moulton himself used this figure when he introduced the Moulton plane in 1902.)

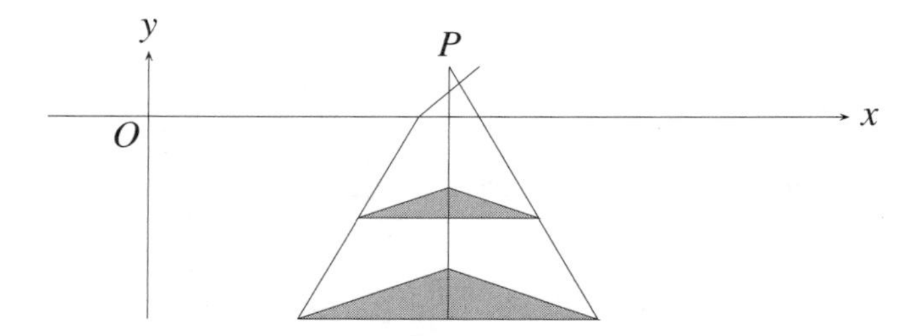

Figure 6.16: The converse Desargues theorem fails in the Moulton plane

6.3.1 Explain why Figure 6.16 shows the failure of the converse Desargues theorem in the Moulton plane.

6.3.2 Formulate a converse to the little Desargues theorem, and show that it follows from the little Desargues theorem.

6.3.3 Show that the converse little Desargues theorem implies a "little scissors theorem" in which the quadrilaterals have their vertices on parallel lines.

6.3.4 Design a figure that directly shows the failure of the little scissors theorem in the Moulton plane.

6.4 Projective arithmetic

If we choose any two lines in a projective plane as the x- and y-axes, we can add and multiply any points on the x-axis by certain constructions. The constructions resemble constructions of Euclidean geometry, but they use straightedge only, so they make sense in projective geometry. To keep them simple, we use lines we call "parallel," but this merely means lines meeting on a designated "line at infinity." The real difficulty is that the construction of $a+b$, for example, is different from the construction of $b+a$, so it is a "coincidence" if $a+b=b+a$. Similarly, it is a "coincidence" if $ab=ba$, or if any other law of algebra holds. Fortunately, we can show that the required coincidences actually occur, because they are implied by certain geometric coincidences, namely, the Pappus and Desargues theorems.

Addition

To construct the sum $a+b$ of points a and b on the x-axis, we take any line $\mathscr{L}$ parallel to the x-axis and construct the lines shown in Figure 6.17:

1. A line from a to the point where $\mathscr{L}$ meets the y-axis.
2. A line from b parallel to the y-axis.
3. A parallel to the first line through the intersection of the second line and $\mathscr{L}$.

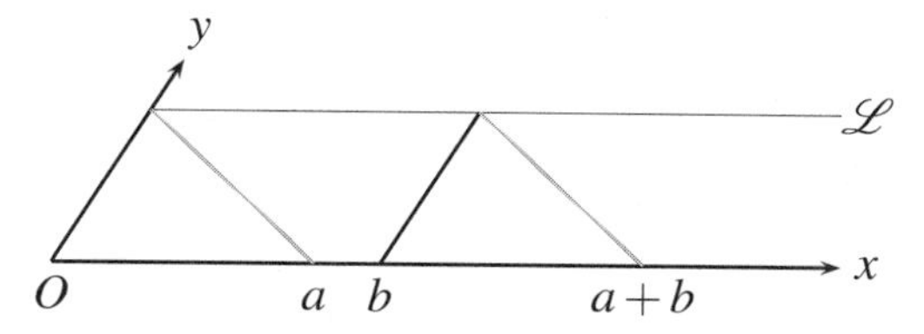

Figure 6.17: Construction of the sum

This construction is similar in spirit to the construction of the sum in Section 1.1. There we "copied a length" by moving it from one place to another by a compass. The spirit of the compass remains in the projective construction: the black line and the gray line form a "compass" that "copies" the length Oa to the point b.

We need the line $\mathscr{L}$ to construct $a+b$, but we get the same point $a+b$ from any other line $\mathscr{L}'$ parallel to the x-axis. This coincidence follows from the little Desargues theorem as shown in Figure 6.18.

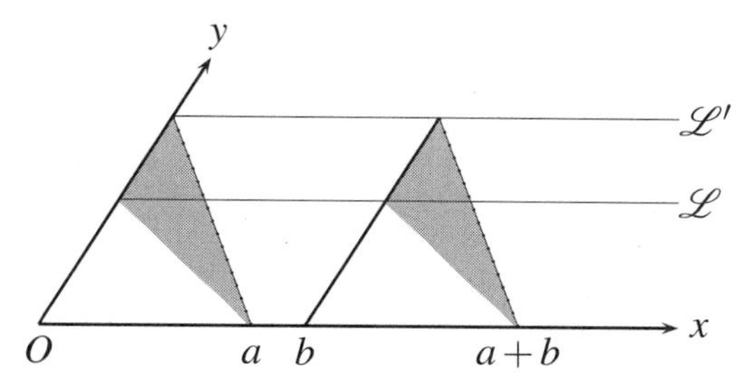

Figure 6.18: Why the sum is independent of the choice of $\mathscr{L}$

The black sides of the solid triangles are parallel by construction, as are the gray sides, one of which ends at the point $a+b$ constructed from $\mathscr{L}$. Then it follows from the little Desargues theorem that the dotted sides are also parallel, and one of them ends at the point $a+b$ constructed from $\mathscr{L}'$. Hence, the same point $a+b$ is constructed from both $\mathscr{L}$ and $\mathscr{L}'$.

Multiplication

To construct the product ab of two points a and b on the x-axis, we first need to choose a point $\neq O$ on the x-axis to be 1. We also choose a point $\neq O$ to be the 1 on the y-axis. The point ab is constructed by drawing the black and gray lines from 1 and a on the x-axis to 1 on the y-axis, and then drawing their parallels as shown in Figure 6.19. This construction is the projective version of "multiplication by a" done in Section 1.4.

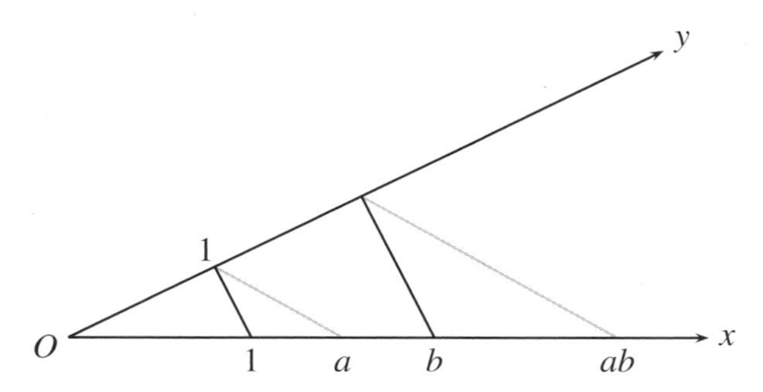

Figure 6.19: Construction of the product ab

Choosing the 1 on the x-axis means choosing a unit of length on the x-axis, so the position of ab definitely depends on it. For example, $ab = b$ if $a = 1$ but $ab \neq b$ if $a \neq 1$. However, the position of ab does not depend on the choice of 1 on the y-axis, as the scissors theorem shows (Figure 6.20).

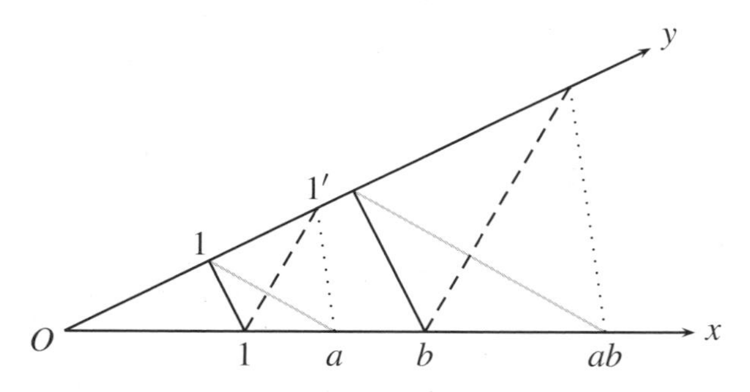

Figure 6.20: Why the product is independent of the 1 on the y-axis

If we choose $1'$ instead of 1 to construct ab, the path from b to ab follows the dashed and the dotted line instead of the black and the gray line. But it ends in the same place, because the dotted line to ab is parallel to the dotted line to a, by the scissors theorem.

Interchangeability of the axes

Once we have chosen points called 1 on both the x- and y-axes, it is natural to let each point a on the x-axis correspond to the point on the y-axis obtained by drawing the line through a parallel to the line through the points 1 on both axes (Figure 6.21).

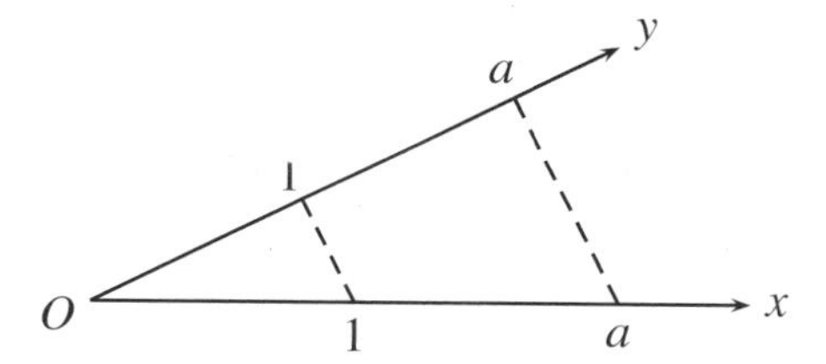

Figure 6.21: Corresponding points

It is also natural to define sum and product on the y-axis by constructions like those on the x-axis. But then the question arises: Do the y-axis sum and product correspond to the x-axis sum and product?

To show that *sums correspond*, we need to construct $a+b$ on the x-axis, and then show that the corresponding point $a+b$ on the y-axis is the y-axis sum of the y-axis a and b. Figure 6.22 shows how this construction is done.

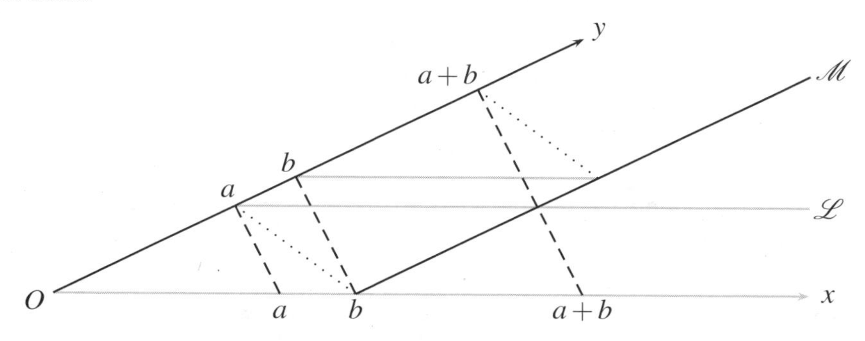

Figure 6.22: Corresponding sums

We construct $a+b$ on the x-axis using the line $\mathscr{L}$ through a on the y-axis. That is, draw the line $\mathscr{M}$ through b on the x-axis and parallel to the y-axis, and then draw the line (dashed) from the intersection of $\mathscr{M}$ and $\mathscr{L}$ parallel to the line from a on the x-axis to the intersection of $\mathscr{L}$ and the y-axis. This dashed line meets the x-axis at $a+b$, and (because it is parallel to the line from a to a) it also meets the y-axis at $a+b$.

Now we construct $a+b$ on the y-axis using the line $\mathscr{M}$ (as on the x-axis, the sum does not depend on line chosen, as long as it is parallel to the y-axis). That is, draw the line through b on the y-axis parallel to the x-axis, and then draw the line (dotted) parallel to the line from a on the y-axis to b on the x-axis (because this b is the intersection of the x-axis with $\mathscr{M}$).

But then, as is clear from Figure 6.22, we have a Pappus configuration of gray, dashed, and dotted lines between the y-axis and $\mathscr{M}$, hence the dotted line (leading to the y-axis sum) and the dashed line (leading to the point corresponding to the x-axis sum) end at the same point, as required. □

To show that *products correspond*, we use the scissors theorem from Section 6.3. Figure 6.23 shows the corresponding points 1, a, b, and ab on both axes. The gray lines construct ab on the x-axis, and the dotted lines construct ab on the y-axis.

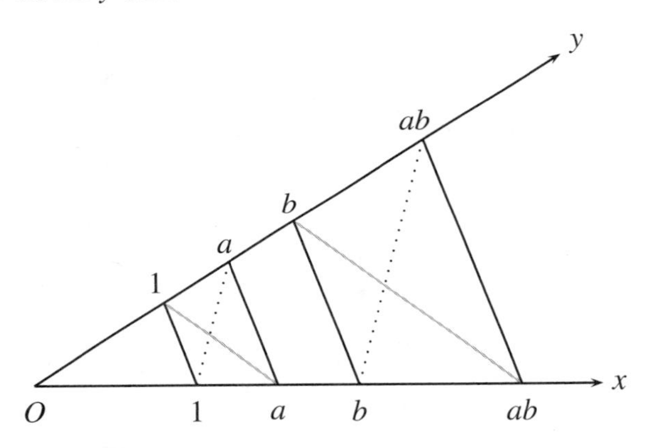

Figure 6.23: Construction of the product ab on both axes

It follows from the scissors theorem that the dotted line on the right ends at the same point as the black line from ab on the x-axis parallel to the lines connecting the corresponding points a and the corresponding points b. Hence, the product of a and b on the y-axis (at the end of the dotted line) is indeed the point corresponding to ab on the x-axis. □

Exercises

These definitions of sum and product lead immediately to some of the simpler laws of algebra, namely, those concerned with the behavior of 0 and 1. The complete list of algebraic laws is given in the Section 6.5.

6.4.1 Show that $a+O=a$ for any a, so O functions as the zero on the x-axis.

6.4.2 Show that, for any a, there is a point b that serves as $-a$; that is, $a+b=O$. (Warning: Do not be tempted to use measurement to find b. Work backward from $O=a+b$, reversing the construction of the sum.)

6.4.3 Show that $a1=a$ for any a.

6.4.4 Show that, for any $a \neq O$, there is a b that serves as a^{-1}; that is, $ab=1$. (Again, do the construction of the product in reverse.)

You will notice that we have not attempted to define sums or products involving the point at infinity ∞ on the x-axis.

6.4.5 What happens when we try to construct $a+\infty$?

6.4.6 What is $-\infty$?

You should find that the answers to Exercises 6.4.5 and 6.4.6 are incompatible with ordinary arithmetic. This is why we do not include ∞ among the points we add and multiply.

6.5 The field axioms

In calculating with numbers, and particularly in calculating with symbols ("algebra"), we assume several things: that there are particular numbers 0 and 1; that each number a has a *additive inverse*, $-a$; that each number $a \neq 0$ has a *reciprocal*, a^{-1}; and that the following *field axioms* hold. (We introduced these in the discussion of vector spaces in Section 4.8.)

$$\begin{array}{rrr}
a+b=b+a, & ab=ba & \text{(commutative laws)} \\
a+(b+c)=(a+b)+c, & a(bc)=(ab)c & \text{(associative laws)} \\
a+0=a, & a1=a & \text{(identity laws)} \\
a+(-a)=0, & aa^{-1}=1 & \text{(inverse laws)} \\
\multicolumn{2}{c}{a(b+c)=ab+ac} & \text{(distributive law)}
\end{array}$$

We generally use these laws unconsciously. They are used so often, and they are so obviously true of numbers, that we do not notice them. But for the projective sum and product of points, they are *not* obviously true. It is not even clear that $a+b=b+a$, because the construction of $a+b$ is different from the construction of $b+a$. It is truly a *coincidence* that $a+b=b+a$ in projective geometry, the result of a geometric coincidence of the type discussed in Section 6.2.

In this chapter, we show that just two coincidences—the theorems of Pappus and Desargues—imply all nine field axioms. In fact, it is known that Pappus alone is sufficient, because it implies Desargues. We do not prove this fact here, partly because it is difficult, and partly because the Desargues theorem itself is interesting: It implies all the field axioms except $ab = ba$. Thus, the theorems of Pappus and Desargues have *algebraic content* that can be measured accurately by the field axioms they imply. Pappus implies all nine, and Desargues only eight—all but $ab = ba$.

Proof of the commutative laws

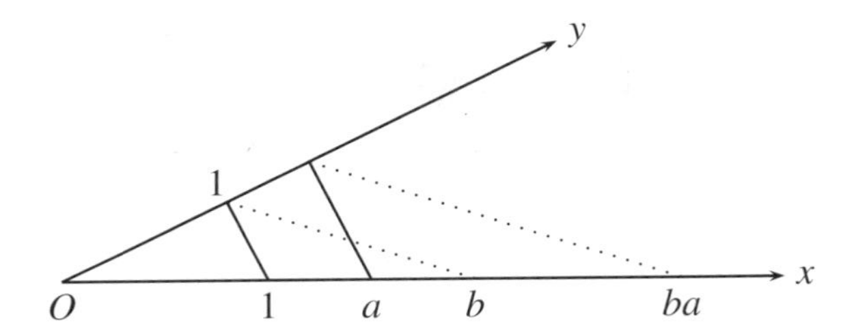

Figure 6.24: Construction of the product ba

We begin with the law $ab = ba$, which is the most important consequence of the Pappus theorem. Figure 6.24 shows the construction of ba from a and b, lying at the end of the second dotted line. It is different from the construction of ab, and Figure 6.25 shows the constructions of both ab and ba on the same diagram.

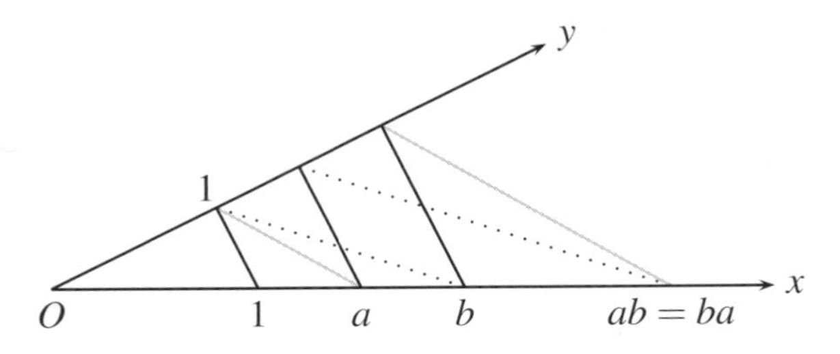

Figure 6.25: Construction of both ab and ba

Then $ab = ba$ because the gray and dotted lines end at the same place, by the Pappus theorem. The Pappus configuration in Figure 6.25 consists of all the lines except the line joining 1 on the x-axis to 1 on the y-axis. □

There is a similar proof that $a+b=b+a$. Remember from Section 6.4 that $a+b$ is the result of attaching the segment Oa at b. Thus, $b+a$ is the result of attaching Ob at a, which is different from the construction of $a+b$. Looking at both constructions together (Figure 6.26), we see that the gray line leads to $a+b$ and the dotted line leads to $b+a$. However, both of these lines end at the same point, thanks to the Pappus theorem.

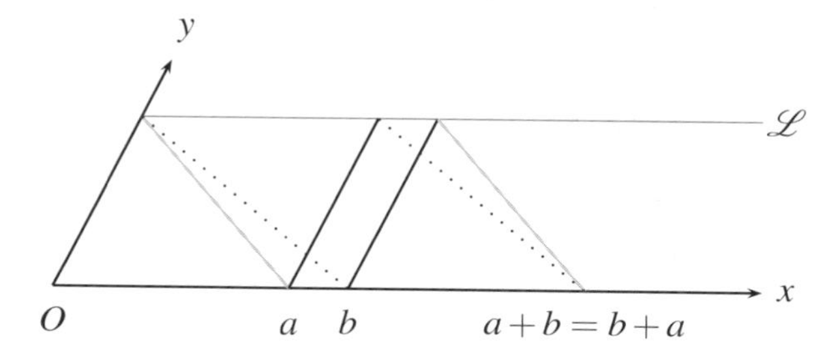

Figure 6.26: Construction of both $a+b$ and $b+a$

Exercises

6.5.1 Look back to the vector proof of the Pappus theorem in Section 4.2, and point out where it uses the assumption $ab=ba$.

The Pappus configuration that proves $a+b=b+a$ is actually a special one, because the vertices of the hexagon lie on parallel lines. The same special configuration also occurs in the proof in Section 6.4 that sums correspond on the x- and y-axes.

The special configuration corresponds to a special Pappus theorem, sometimes called the "little Pappus theorem." It is usually stated without mention of parallel lines; in which case, one has to talk about opposite sides of the hexagon meeting on a line $\mathcal{L}$.

6.5.2 Given that the assumptions of the little Pappus theorem are a hexagon with vertices on two lines that meet at a point P, and two pairs of opposite sides meeting on a line $\mathcal{L}$ that goes through P, what is the conclusion?

It is known that the little Desargues theorem implies the little Pappus theorem; a proof is in *Fundamentals of Mathematics, II* by Behnke *et al.*, p. 70. Thus, the results deduced here from the little Pappus theorem can also be deduced from the little Desargues theorem (although generally not as easily).

6.6 The associative laws

First we look at the associative law of addition, $a+(b+c)=(a+b)+c$. Figure 6.27 shows the construction of $a+(b+c)$. We have to construct $b+c$ from b and c first, and then add a as was done in Figure 6.17. Next we have to construct $(a+b)+c$, which means constructing $a+b$ first and then adding it to c as shown in Figure 6.28.

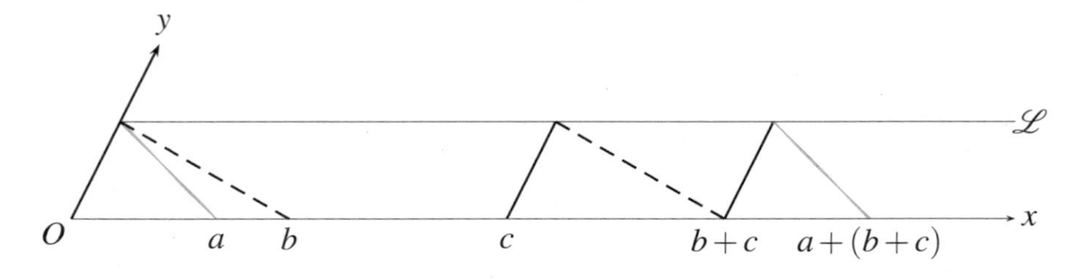

Figure 6.27: Construction of $a+(b+c)$

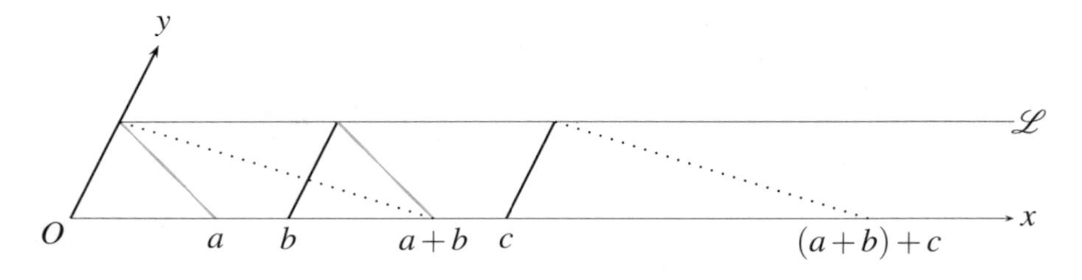

Figure 6.28: Construction of $(a+b)+c$

Figure 6.29 shows both Figures 6.27 and 6.28 on the same diagram. Here we need Desargues or, more precisely, the scissors theorem.

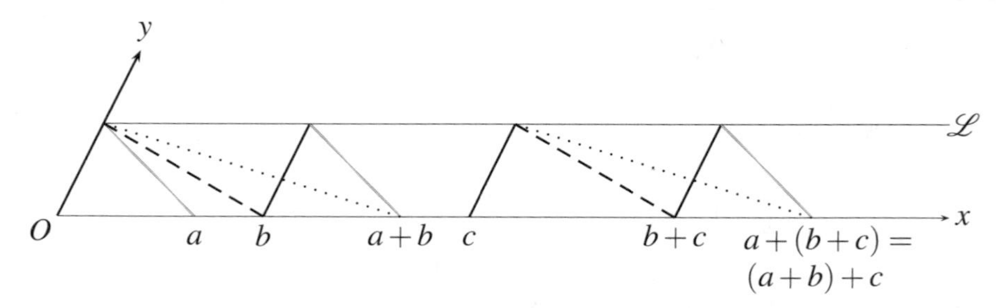

Figure 6.29: Why $a+(b+c)$ and $(a+b)+c$ coincide

One can clearly see two pairs of scissors, each consisting of a dashed line, a dotted line, a black line, and a gray line. In the scissors on the right, the gray line ends at $a+(b+c)$ and the dotted line at $(a+b)+c$. But the ends of these lines coincide, by the scissors theorem. Hence $a+(b+c) = (a+b)+c$. □

Because the scissors in this proof lie between parallel lines, we need only the little scissors theorem (and hence only the little Desargues theorem, by the remark in the previous exercise set).

Next we consider the associative law of multiplication, $a(bc) = (ab)c$. The diagram (Figure 6.30) is similar, except that the pairs of scissors lie between nonparallel lines (the x- and y-axes), so now we need the full Desargues theorem.

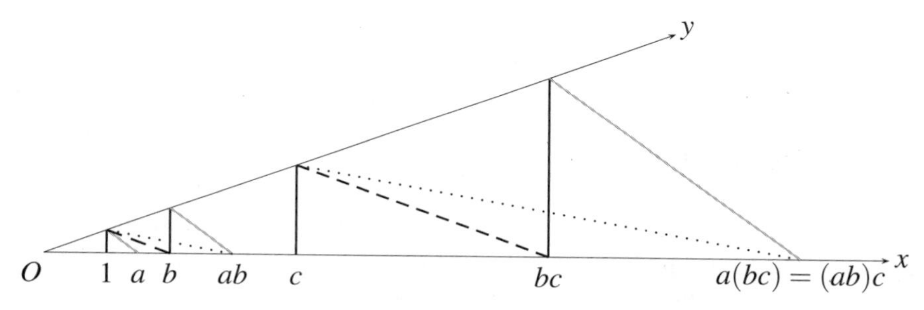

Figure 6.30: Why $a(bc)$ and $(ab)c$ coincide

The gray line ends at $a(bc)$ and the dotted line ends at $(ab)c$. But the ends of these lines coincide, by the scissors theorem, so $a(bc) = (ab)c$. □

Exercises

There is an algebraic system that satisfies all of the field axioms except the commutative law of multiplication. It is called the *quaternions* and is denoted by $\mathbb{H}$, after Sir William Rowan Hamilton, who discovered the quaternions in 1843.

In 1845, Arthur Cayley showed that the quaternions could be defined as 2×2 complex matrices of the form

$$\mathbf{q} = \begin{pmatrix} a+ib & c+id \\ -c+id & a-ib \end{pmatrix}.$$

Most of their properties follow from general properties of matrices. In fact, all the laws of algebra are immediate except the existence of $\mathbf{q}^{-1}$ and commutative multiplication.

6.6.1 Show that $\mathbf{q}$ has determinant $a^2+b^2+c^2+d^2$ and hence that $\mathbf{q}^{-1}$ exists for any nonzero quaternion $\mathbf{q}$.

6.6.2 Find specific quaternions $\mathbf{s}$ and $\mathbf{t}$ such that $\mathbf{st} \neq \mathbf{ts}$.

We can write any quaternion as $\mathbf{q} = a\mathbf{1} + b\mathbf{i} + c\mathbf{j} + d\mathbf{k}$, where

$$\mathbf{1} = \begin{pmatrix} 1 & 0 \\ 0 & 1 \end{pmatrix}, \quad \mathbf{i} = \begin{pmatrix} i & 0 \\ 0 & -i \end{pmatrix}, \quad \mathbf{j} = \begin{pmatrix} 0 & 1 \\ -1 & 0 \end{pmatrix}, \quad \mathbf{k} = \begin{pmatrix} 0 & i \\ i & 0 \end{pmatrix}.$$

6.6.3 Verify that $\mathbf{i}^2 = \mathbf{j}^2 = \mathbf{k}^2 = \mathbf{ijk} = -\mathbf{1}$ (This is Hamilton's description of the quaternions).

It is possible to define the *quaternion projective plane* $\mathbb{HP}^2$ using quaternion coordinates. $\mathbb{HP}^2$ satisfies the Desargues theorem because it is possible to do the necessary calculations without using commutative multiplication. But it does *not* satisfy the Pappus theorem, because this implies commutative multiplication for $\mathbb{H}$. $\mathbb{HP}^2$ is therefore a *non-Pappian* plane—probably the most natural example.

6.7 The distributive law

To prove the distributive law $a(b+c) = ab+ac$, we take advantage of the ability to do addition and multiplication on both axes. We construct $b+c$ from b and c on the x-axis, and then map b, c, and $b+c$ to ab, ac, and $a(b+c)$ on the y-axis by lines parallel to the line from 1 on the x-axis to a on the y-axis. Then we use addition on the y-axis to construct $ab+ac$ there, and finally, use the Pappus theorem to show that $ab+ac$ and $a(b+c)$ are the same point. Here are the details.

First, observe that we can map any b on the x-axis to ab on the y-axis by sending it along a line parallel to the line from 1 on the x-axis to a on the y-axis (Figure 6.31).

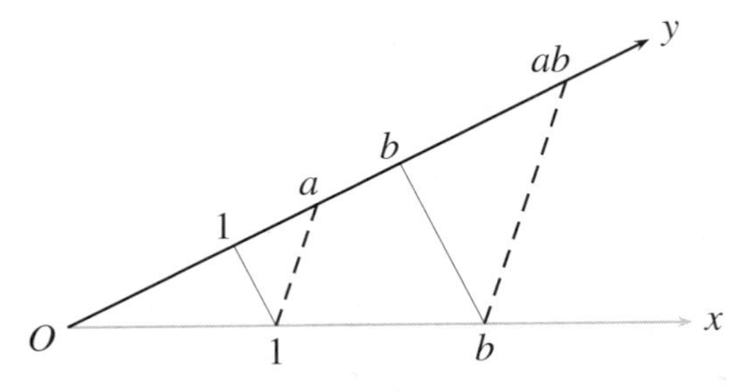

Figure 6.31: Multiplication via parallels

This is the same as constructing ab from a and b on the y-axis, because the line from b to b is parallel to the line from 1 to 1, as required by the definition of multiplication.

Next we add b and c on the x-axis, using a special choice of line $\mathscr{L}$: the parallel through ab on the y-axis. We also connect b, c, and $b+c$, respectively, to ab, ac, and $a(b+c)$ on the y-axis by parallel lines, shown dashed in Figure 6.32. The line through c that constructs $b+c$, namely the parallel $\mathscr{M}$ to the y-axis, is used in turn to add ab and ac on the y-axis.

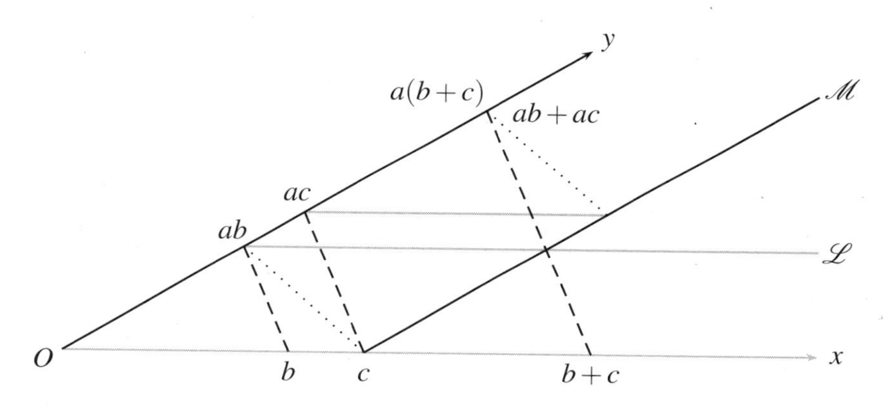

Figure 6.32: Why $a(b+c) = ab + ac$

This figure has the same structure as Figure 6.22; only the labels have changed. Now the dashed line ends at $a(b+c)$, and the dotted line ends at $ab+ac$. But again the endpoints coincide by the theorem of Pappus, and so $a(b+c) = ab+ac$. □

Exercises

We need not prove the other distributive law, $(b+c)a = ba+bc$, because we are assuming Pappus, so multiplication is commutative.

6.7.1 Explain in this case why $a(b+c) = ab+ac$ implies $(b+c)a = ba+bc$.

However, in some important systems with noncommutative multiplication, both distributive laws remain valid.

6.7.2 Explain why both distributive laws are valid for the quaternions.

6.7.3 More generally, show that both distributive laws are valid for matrices.

6.8 Discussion

The idea of developing projective geometry without the use of numbers comes from the German mathematician Christian von Staudt in 1847. His compatriots Hermann Wiener and David Hilbert took the idea further in the 1890s, and it reached a high point with the publication of Hilbert's book, *Grundlagen der Geometrie* (Foundations of geometry), in 1899. It was Hilbert who first established a clear correlation between geometric and algebraic structure:

- Pappus with commutative multiplication,
- Desargues with associative multiplication.

The correlation is significant because some important algebraic systems satisfy all the field axioms except commutative multiplication. The best-known example is the quaternions, which has been known since 1843, but, for some reason, Hilbert did not mention it. To construct a non-Pappian plane, he created a rather artificial noncommutative coordinate system.

It is perhaps a lucky accident of history that Hilbert discovered the role of the Desargues theorem at all. He was forced to use it because, in 1899, it was still not known that Pappus implies Desargues. This implication was first proved by Gerhard Hessenberg in 1904. Even then the proof was faulty, and the mistake was not corrected until years later.

The whole circle of ideas was neatly tied up by yet another German mathematician, Ruth Moufang, in 1930. She found that the *little* Desargues theorem also has algebraic significance. In a projective plane satisfying the little Desargues theorem, with addition and multiplication defined as in Section 6.4, one can prove all the field axioms *except* commutativity and associativity. One can even prove a partial associativity law called *cancellation* or *alternativity*:

$$a^{-1}(ab) = b = (ba)a^{-1} \qquad \text{(alternativity)}$$

The commutative, associative, and alternative laws are beautifully exemplified by the possible multiplication operations that can be defined "reasonably" on the Euclidean spaces $\mathbb{R}^n$. ("Reasonably" means respecting at least the dimension of the space. For more on the problem of generalizing the idea of number to n dimensions, see the book *Numbers* by D. Ebbinghaus *et al.*)

- Commutative multiplication is possible only on $\mathbb{R}^1$ and $\mathbb{R}^2$, and it yields the number systems $\mathbb{R}$ and $\mathbb{C}$.

- Associative, but noncommutative, multiplication is possible only on $\mathbb{R}^4$, and it yields the quaternions $\mathbb{H}$.

- Alternative, but nonassociative, multiplication is possible only on $\mathbb{R}^8$, and it yields a system called the *octonions* $\mathbb{O}$. The octonions were discovered by a friend of Hamilton called John Graves, in 1843, and they were discovered independently by Cayley in 1845.

Ruth Moufang was the first to recognize the importance of quaternions and octonions in projective geometry. She pointed out the quaternion projective plane, as a natural example of a non-Pappian plane, and was the first to discuss the *octonion projective plane* $\mathbb{OP}^2$. $\mathbb{OP}^2$ is the most natural example of a plane that satisfies little Desargues but not Desargues.

In Section 5.4, we sketched the construction of the *real projective space* $\mathbb{RP}^3$ by means of homogeneous coordinates. This idea is easily generalized to obtain the n-dimensional real projective space $\mathbb{RP}^n$, and one can obtain $\mathbb{CP}^n$ and $\mathbb{HP}^n$ in precisely the same way. Surprisingly, the idea does *not* work for the octonions. The only octonion projective spaces are the octonion projective line $\mathbb{OP}^1 = \mathbb{O} \cup \{\infty\}$ and the octonion projective plane $\mathbb{OP}^2$ discovered by Moufang.

The reason for the nonexistence of $\mathbb{OP}^3$ is extremely interesting and has to do with the nature of the Desargues theorem in three dimensions. Remember that the Desargues theorem assumes a pair of triangles in perspective and concludes that the intersections of corresponding sides lie on a line. We know (because of the example of the Moulton plane) that the conclusion does not follow by basic incidence properties of points and lines. But *if the triangles lie in three-dimensional space, the conclusion follows by basic incidence properties of points, lines, and planes.*

The spatial Desargues theorem is clear from a picture that emphasizes the placement of the triangles in three dimensions, such as Figure 6.33. The planes containing the two triangles meet in a line $\mathscr{L}$, where the pairs of corresponding sides necessarily meet also.

The argument is a little trickier if the two triangles lie in the same plane. But, *provided the plane lies in a projective space*, it can be carried out (one shows that the planar configuration is a "shadow" of a spatial configuration).

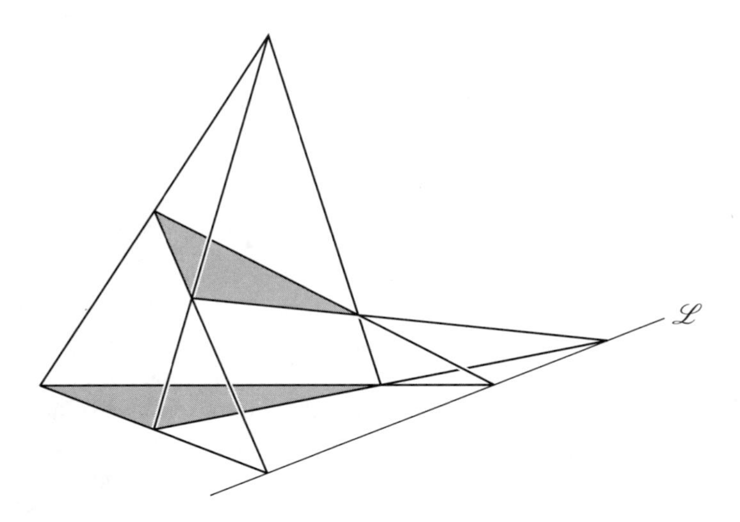

Figure 6.33: The spatial Desargues configuration

Thus, *the Desargues theorem holds in any projective space of at least three dimensions.* This is why $\mathbb{OP}^3$ cannot exist. If it did, the Desargues theorem would hold in it, and we could then show that $\mathbb{O}$ is associative—which it is not. Q. E. D.

7

Transformations

PREVIEW

In this book, we have seen at least three geometries: Euclidean, vector, and projective. In Euclidean geometry, the basic concept is length, but angle and straightness derive from it. In vector geometry, the basic concepts are vector sums and scalar multiples, but from these we derive others, such as midpoints of line segments. Finally, in projective geometry, the basic concept is straightness. Length and angle have no meaning, but a certain combination of lengths—the *cross-ratio*—is meaningful because it is unchanged by projection.

We found the cross-ratio as an *invariant of projective transformations*. The concept of length was not discovered this way, but nevertheless, it *is* an invariant of certain transformations. It is an invariant of the *isometries*, for the simple reason that isometries are *defined* to be transformations of the plane that preserve length.

These examples are two among many that suggest *geometry is the study of invariants of groups of transformations*. This definition of geometry was first proposed by the German mathematician Felix Klein in 1872. Klein's concept of geometry is perhaps still not broad enough, but it does cover the geometry in this book.

In this chapter we look again at Euclidean, vector, and projective geometry from Klein's viewpoint, first explaining precisely what "transformation" and "group" mean. It turns out that the appropriate transformations for projective geometry are *linear*, and that linear transformations also play an important role in Euclidean and vector geometry.

Linear transformations also pave the way for *hyperbolic geometry*, a new "non-Euclidean" geometry that we study in Chapter 8.

7.1 The group of isometries of the plane

In Chapter 3, we took up Euclid's idea of "moving" geometric figures, and we made it precise in the concept of an *isometry* of the plane $\mathbb{R}^2$. An isometry is defined to be a function $f : \mathbb{R}^2 \to \mathbb{R}^2$ that preserves distance; that is,

$$|f(P_1)f(P_2)| = |P_1P_2| \quad \text{for any points } P_1, P_2, \in \mathbb{R}^2,$$

where $|P_1P_2| = \sqrt{(x_2-x_1)^2+(y_2-y_1)^2}$ denotes the distance between the points $P_1 = (x_1, y_1)$ and $P_2 = (x_2, y_2)$.

It follows immediately from this definition that, when f and g are isometries, so is their *composite* or *product* fg (the result of applying g, then f). Namely,

$$\begin{aligned} |f(g(P_1))f(g(P_2))| &= |g(P_1)g(P_2)| \quad \text{because } f \text{ is an isometry} \\ &= |P_1P_2| \quad \text{because } g \text{ is an isometry.} \end{aligned}$$

What is less obvious is that *any isometry f has an inverse, f^{-1}, which is also an isometry.* To prove this fact, we use the result from Section 3.7 that any isometry of $\mathbb{R}^2$ is the product of one, two, or three reflections.

First suppose that $f = r_1r_2r_3$, where r_1, r_2, and r_3 are reflections. Then, because a reflection composed with itself is the identity function, we find

$$\begin{aligned} fr_3r_2r_1 &= r_1r_2r_3r_3r_2r_1 \\ &= r_1r_2r_2r_1 \quad \text{because } r_3r_3 \text{ is the identity function} \\ &= r_1r_1 \quad \text{because } r_2r_2 \text{ is the identity function} \\ &= \text{identity function,} \end{aligned}$$

and therefore, $r_3r_2r_1 = f^{-1}$. This calculation also shows that f^{-1} is an isometry, because it is a product of reflections. The proof is similar (but shorter) when f is the product of one or two reflections.

These properties of isometries are characteristic of a *group of transformations*. A *transformation* of a set S is a function from S to S, and a collection G of transformations forms a *group* if it has the two properties:

- If f and g are in G, then so is fg.
- If f is in G, then so is its inverse, f^{-1}.

It follows that G includes the identity function ff^{-1}, which can be written as 1. This notation is natural when we write the composite of two functions f, g as the "product" fg.

What is a geometry?

In 1872, the German mathematician Felix Klein pointed out that various kinds of geometry go with various groups of transformations. For example, the Euclidean geometry of $\mathbb{R}^2$ goes with the group of isometries of $\mathbb{R}^2$. The meaningful concepts of the geometry correspond to properties that are left *unchanged* by transformations in the group. Isometries of $\mathbb{R}^2$ leave distance or length unchanged, so distance is a meaningful concept of Euclidean geometry. It is called an *invariant* of the isometry group of $\mathbb{R}^2$. This invariance is no surprise, because isometries are *defined* as the transformations that preserve distance.

However, it is interesting that other things are also invariant under isometries, such as straightness of lines and circularity of circles. It is not entirely obvious that a length-preserving transformation preserves straightness, but it can be proved by showing first that any reflection preserves straightness, and then using the theorem of Section 3.7, that any isometry is a product of reflections.

An example of a concept without meaning in Euclidean geometry is "being vertical," because a vertical line can be transformed to a nonvertical line by an isometry (for example, by a rotation). We can do without the concept of "vertical" in geometry because we have the concept of "being relatively vertical," that is, perpendicular. A concept that is harder to do without is "clockwise order on the circle." This concept has no meaning in Euclidean geometry because the points $A=(-1,0)$, $B=(0,1)$, $C=(1,0)$, and $D=(0,-1)$ have clockwise order on the circle, but their respective reflections in the x-axis do not.

However, we can define *oriented Euclidean geometry*, in which clockwise order is meaningful, by using a smaller group of transformations. Instead of the group $\mathrm{Isom}(\mathbb{R}^2)$ of all isometries of $\mathbb{R}^2$, take $\mathrm{Isom}^+(\mathbb{R}^2)$, each member of which is the product of an *even* number of reflections. $\mathrm{Isom}^+(\mathbb{R}^2)$ is a group because

- If f and g are products of an even number of reflections, so is fg.
- If $f=r_1r_2\cdots r_{2n}$ is the product of an even number of reflections, then so is f^{-1}. In fact, $f^{-1}=r_{2n}\cdots r_2r_1$, by the argument used above to invert the product of any number of reflections.

And any transformation in $\mathrm{Isom}^+(\mathbb{R}^2)$ preserves clockwise order because any product of two reflections does: The first reflection reverses the order, and then the second restores it.

This example shows how a geometry of $\mathbb{R}^2$ with more concepts comes from a group with fewer transformations. In $\mathbb{R}^3$, one has the concept of "handedness"—which distinguishes the right hand from the left—which is not preserved by all isometries of $\mathbb{R}^3$. However, it is preserved by products of an even number of reflections in planes. Thus, the geometry of $\mathrm{Isom}(\mathbb{R}^3)$ does not have the concept of handedness, but the geometry of $\mathrm{Isom}^+(\mathbb{R}^3)$ does. Restricting the transformations to those that preserve *orientation*—as it is generally called—is a common tactic in geometry.

However, the main goal of this book is to show that there are interesting geometries with *fewer* concepts than Euclidean geometry. These geometries are obtained by taking larger groups of transformations, which we study in the remainder of this chapter.

Exercises

7.1.1 Use the results of Section 3.7 to show that each member of $\mathrm{Isom}^+(\mathbb{R}^2)$ is either a translation or a rotation.

7.1.2 Why does an isometry map any circle to a circle?

We took care to write the inverse of the isometry $r_1r_2r_3$ as $r_3r_2r_1$ because only this ordering of terms will always give the correct result.

7.1.3 Give an example of two reflections r_1 and r_2 such that $r_1r_2 \neq r_2r_1$.

7.2 Vector transformations

In Chapter 4, we viewed the plane $\mathbb{R}^2$ as a *real vector space*, by considering its points to be *vectors* that can be added and multiplied by scalars. If $\mathbf{u} = (u_1, u_2)$ and $\mathbf{v} = (v_1, v_2)$, we defined the *sum* of $\mathbf{u}$ and $\mathbf{v}$ by

$$\mathbf{u} + \mathbf{v} = (u_1 + v_1, u_2 + v_2)$$

and the *scalar multiple* $a\mathbf{u}$ of $\mathbf{u}$ by a real number a by

$$a\mathbf{u} = (au_1, au_2).$$

A transformation f of $\mathbb{R}^2$ *preserves* these two operations on vectors if

$$f(\mathbf{u} + \mathbf{v}) = f(\mathbf{u}) + f(\mathbf{v}) \quad \text{and} \quad f(a\mathbf{u}) = af(\mathbf{u}), \tag{*}$$

and such a transformation is called *linear*.

One reason for calling the transformation "linear" is that it preserves straightness of lines. A straight line is a set of points of the form $\mathbf{a}+t\mathbf{b}$, where $\mathbf{a}$ and $\mathbf{b}$ are constant vectors and t runs through the real numbers. Figure 7.1 shows the role of the vectors $\mathbf{a}$ and $\mathbf{b}$: $\mathbf{a}$ is one point on the line, and $\mathbf{b}$ gives the direction of the line.

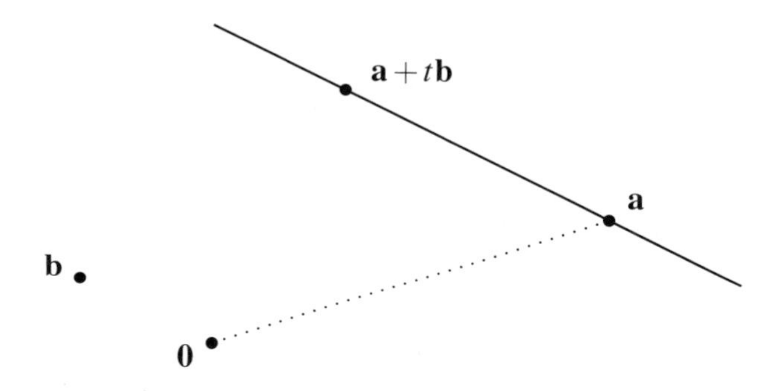

Figure 7.1: Points on a line

If we apply a linear transformation f to this set of points, we get the set of points $f(\mathbf{a}+t\mathbf{b})$. And by the linearity conditions (*), this set consists of points of the form $f(\mathbf{a})+tf(\mathbf{b})$, which is another straight line: $f(\mathbf{a})$ is one point on it, and $f(\mathbf{b})$ gives the direction of the line.

It follows from this calculation that, if $\mathscr{L}_1$ and $\mathscr{L}_2$ are two lines with direction $\mathbf{b}$ and f is a linear transformation, then $f(\mathscr{L}_1)$ and $f(\mathscr{L}_2)$ are two lines with direction $f(\mathbf{b})$. In other words, a linear transformation also preserves parallels.

Matrix representation

Another consequence of the linearity conditions (*) is that each linear transformation f of $\mathbb{R}^2$ can be specified by four real numbers a,b,c,d: any point (x,y) of $\mathbb{R}^2$ is sent by f to the point $(ax+by,cx+dy)$.

Certainly, there are numbers a,b,c,d that give the particular values

$$f((1,0))=(a,c) \quad \text{and} \quad f((0,1))=(b,d).$$

But the value of $f((x,y))$ follows from these particular values by linearity:

$$(x,y)=x(1,0)+y(0,1),$$

and therefore,

$$\begin{aligned} f((x,y)) &= f(x(1,0)+y(0,1)) \\ &= xf((1,0))+yf((0,1)) \\ &= x(a,c)+y(b,d) \\ &= (ax+by, cx+dy). \end{aligned}$$

The linear transformation $(x,y) \mapsto (ax+by, cx+dy)$ is usually represented by the *matrix*

$$M = \begin{pmatrix} a & b \\ c & d \end{pmatrix}, \quad \text{where} \quad a,b,c,d \in \mathbb{R}.$$

To find where $(x,y) \in \mathbb{R}^2$ is sent by f, one writes it as the "column vector" $\begin{pmatrix} x \\ y \end{pmatrix}$ and multiplies this column on the left by M according to the matrix product rule:

$$\begin{pmatrix} a & b \\ c & d \end{pmatrix} \begin{pmatrix} x \\ y \end{pmatrix} = \begin{pmatrix} ax+by \\ cx+dy \end{pmatrix}.$$

The main advantage of the matrix notation is that it gives the product of two linear transformations, first $(x,y) \mapsto (a_2x+b_2y, c_2x+d_2y)$ and then $(x,y) \mapsto (a_1x+b_1y, c_1x+d_1y)$, by the matrix product rule:

$$\begin{pmatrix} a_1 & b_1 \\ c_1 & d_1 \end{pmatrix} \begin{pmatrix} a_2 & b_2 \\ c_2 & d_2 \end{pmatrix} = \begin{pmatrix} a_1a_2+b_1c_2 & a_1b_2+b_1d_2 \\ c_1a_2+d_1c_2 & c_1b_2+d_1d_2 \end{pmatrix}.$$

Matrix notation also exposes the role of the *determinant*, $\det(M)$, which must be nonzero for the linear transformation to have an inverse. If

$$M = \begin{pmatrix} a & b \\ c & d \end{pmatrix}, \quad \text{then} \quad \det(M) = ad - bc,$$

and if $\det(M) \neq 0$, then

$$M^{-1} = \frac{1}{\det M} \begin{pmatrix} d & -b \\ -c & a \end{pmatrix}.$$

Examples of linear transformations

Any 2×2 real matrix M represents a linear transformation, because it follows from the definition of matrix multiplication that

$$M(\mathbf{u}+\mathbf{v}) = M\mathbf{u}+M\mathbf{v} \quad \text{and} \quad M(a\mathbf{u}) = aM\mathbf{u}$$

for any vectors $\mathbf{u}$ and $\mathbf{v}$ (written in column form).

Among the invertible linear transformations are certain isometries, such as rotations and reflections in lines through the origin. Recall from Section 3.6 that a rotation is a transformation of the form

$$(x,y) \mapsto (cx-sy, sx+cy), \quad \text{hence given by the matrix} \quad R = \begin{pmatrix} c & -s \\ s & c \end{pmatrix}.$$

The numbers c and s satisfy $c^2+s^2=1$ (they are actually $\cos\theta$ and $\sin\theta$, where θ is the angle of rotation); hence,

$$\det R = 1 \quad \text{and therefore} \quad R^{-1} = \begin{pmatrix} c & s \\ -s & c \end{pmatrix}.$$

Likewise, reflection in the x-axis is the linear transformation

$$(x,y) \mapsto (x,-y), \quad \text{given by the matrix} \quad \overline{X} = \begin{pmatrix} 1 & 0 \\ 0 & -1 \end{pmatrix}.$$

We can reflect $\mathbb{R}^2$ in any line $\mathscr{L}$ through O with the help of the rotation R that sends the x-axis to $\mathscr{L}$:

- First apply R^{-1} to send $\mathscr{L}$ to the x-axis.
- Then carry out the reflection by applying $\overline{X}$.
- Then send the line of reflection back to $\mathscr{L}$ by applying R.

In other words, to reflect the point $\mathbf{u}$ in $\mathscr{L}$, we find the value of $R\overline{X}R^{-1}\mathbf{u}$. Hence, reflection in $\mathscr{L}$ is represented by the matrix $R\overline{X}R^{-1}$.

Thus, the linear transformations of $\mathbb{R}^2$ include the isometries that are products of reflection on lines through O. But this is not all. An example of a linear transformation that is not an isometry is the *stretch by factor k* in the x-direction,

$$(x,y) \mapsto (kx,y), \quad \text{given by the matrix} \quad S = \begin{pmatrix} k & 0 \\ 0 & 1 \end{pmatrix}.$$

It can be shown that any invertible linear transformation of $\mathbb{R}^2$ is a product of reflections in lines through O and stretches in the x-direction (by factors $k \neq 0$).

Affine transformations

Linear transformations preserve geometrically natural properties such as straightness and parallelism, but they also preserve the origin, which really is not geometrically different from any other point. To abolish the special position of the origin, we allow linear transformations to be composed with translations, obtaining what are called *affine* transformations. If we write an arbitrary linear transformation of (column) vectors $\mathbf{u}$ in the form

$$f(\mathbf{u}) = M\mathbf{u}, \quad \text{where } M \text{ is an invertible matrix,}$$

then an arbitrary affine transformation takes the form

$$g(\mathbf{u}) = M\mathbf{u} + \mathbf{c}, \quad \text{where } \mathbf{c} \text{ is a constant vector.}$$

Because translations preserve everything except position, affine transformations preserve everything that linear transformations do, except position. In effect, they allow any point to become the origin.

The geometry of affine transformations is called *affine geometry*. Its theorems include those in the first few sections of Chapter 4, such as the fact that diagonals of a parallelogram bisect each other, and the concurrence of the medians of a triangle. These theorems belong to affine geometry because they are concerned only with quantities, such as the midpoint of a line segment, that are preserved by affine transformations.

Exercises

7.2.1 Compute MM^{-1} for the general 2×2 matrix $M = \begin{pmatrix} a & b \\ c & d \end{pmatrix}$, and verify that it equals the identity matrix $\begin{pmatrix} 1 & 0 \\ 0 & 1 \end{pmatrix}$.

7.2.2 Write down the matrix for clockwise rotation through angle $\pi/4$.

7.2.3 Write down the matrix for reflection in the line $y = x$, and check that it equals $R\overline{X}R^{-1}$, where R is the matrix for rotation through $\pi/4$ found in Exercise 7.2.2.

7.2.4 The matrix $M = \begin{pmatrix} k & 0 \\ 0 & k \end{pmatrix}$ represents a dilation of the plane by factor k (also known as a *similarity* transformation). Explain geometrically why this transformation is a product of reflections in lines through O and of stretches by factor k in the x-direction.

7.2.5 Show that the midpoint of any line segment is preserved by linear transformations and hence by affine transformations.

7.2.6 More generally, show that the ratio of lengths of any two segments of the same line is preserved by affine transformations.

7.3 Transformations of the projective line

Looking back at our approach to the projective line in Chapter 5, we see that we were following Klein's idea. First we found the transformations of the projective line, and then a quantity that they leave invariant—the cross-ratio. In this section we look more closely at projective transformations, and show that they too can be viewed as linear transformations.

In Sections 5.5 and 5.6, we showed that the transformations of the projective line $\mathbb{R} \cup \{\infty\}$ are precisely the linear fractional functions

$$f(x) = \frac{ax+b}{cx+d} \quad \text{where} \quad ad - bc \neq 0.$$

We did this by showing:

- Any linear fractional function is a product of functions sending x to $x+l$, x to kx, and x to $1/x$, and that each of the latter functions can be realized by projection of one line onto another.
- Conversely, any projection of one line onto another is represented by a linear fractional function of x, with the understanding that $1/0 = \infty$ and $1/\infty = 0$.

In the exercises to Section 5.6 you were asked to show that linear fractional functions $f(x) = \frac{ax+b}{cx+d}$ behave like the matrices $M = \begin{pmatrix} a & b \\ c & d \end{pmatrix}$, by showing that composition of the functions corresponds to multiplication of the corresponding matrices. In this section, we explain the connection by representing mappings of the projective line directly by linear transformations of the plane.

We begin by defining the projective line in the manner of Section 5.4. There we defined the *real projective plane* $\mathbb{RP}^2$. Its "points" are the lines through O in $\mathbb{R}^3$, and its "lines" are the planes through O. Here we need only one projective line, which we can take to be the real projective line $\mathbb{RP}^1$, whose "points" are the lines through O in the ordinary plane $\mathbb{R}^2$.

We label each line through O, if it meets the line $y = 1$, by the x-coordinate s of the point of intersection (Figure 7.2). The single line that does *not* meet $y = 1$, namely, the x-axis, naturally gets the label ∞.

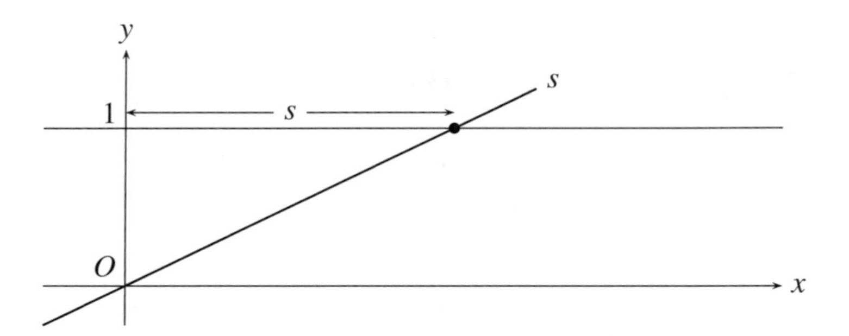

Figure 7.2: Correspondence between lines through O and points on $y = 1$

Figure 7.3 shows some lines through O with their labels.

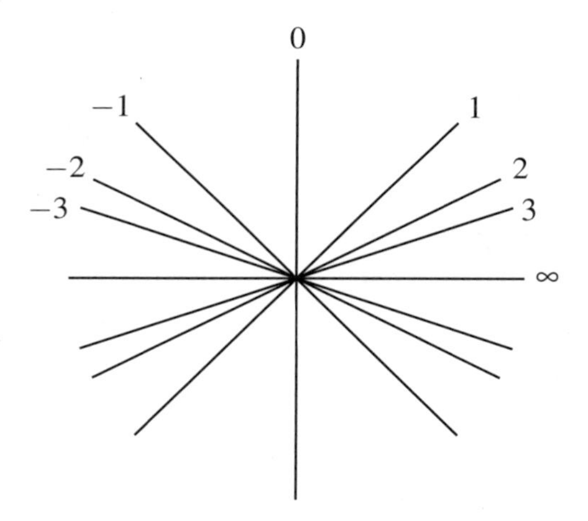

Figure 7.3: Labeling of lines through O

Now a projective map of the ordinary line $y = 1$ sends the point with x-coordinate s to the point with x-coordinate $f(s)$, for some linear fractional function

$$f(s) = \frac{as+b}{cs+d}.$$

This function corresponds to a map of the plane $\mathbb{R}^2$ sending the line with label s to the line with label $\frac{as+b}{cs+d}$. Bearing in mind that the label represents the "reciprocal slope" ("run over rise") of the line, we find that *one such map is the linear map of the plane given by the matrix*

$$M = \begin{pmatrix} a & b \\ c & d \end{pmatrix}.$$

To see why, we apply this linear map to a typical point (sx, x) on the line with label s. We find where M sends it by writing (sx, x) as a column vector and multiplying it on the left by M:

$$\begin{pmatrix} a & b \\ c & d \end{pmatrix} \begin{pmatrix} sx \\ x \end{pmatrix} = \begin{pmatrix} asx + bx \\ csx + dx \end{pmatrix}.$$

The column vector $\begin{pmatrix} asx + bx \\ csx + dx \end{pmatrix}$ represents the point $(asx + bx, csx + dx)$, which lies on the line with reciprocal slope

$$\frac{asx + bx}{csx + dx} = \frac{as + b}{cs + d}.$$

The latter line is therefore independent of x and it is the line with label $\frac{as+b}{cs+d}$. Thus, M maps the line with label s to the line with label $\frac{as+b}{cs+d}$, as required. □

Because a "point" of $\mathbb{RP}^1$ is a whole line through O, we care only that the matrix

$$M = \begin{pmatrix} a & b \\ c & d \end{pmatrix}$$

sends the *line* with label s to the *line* with label $\frac{as+b}{cs+d}$. It does not matter how M moves individual points on the line. It is *not* generally the case that M sends the particular point $(s, 1)$ on the line with label s to the particular point $(\frac{as+b}{cs+d}, 1)$ on the line with label $\frac{as+b}{cs+d}$. Indeed, it is clearly impossible when M represents the map $s \mapsto 1/s$ of $\mathbb{RP}^1$. A matrix M sends each point of $\mathbb{R}^2$ to another point of $\mathbb{R}^2$. Hence it cannot send $(0, 1)$ to $(1/0, 1) = (\infty, 1)$, because the latter is not a point of $\mathbb{R}^2$. However, M *can* send the line with label 0 (the y-axis) to the line with label ∞ (the x-axis), and this is exactly what happens (see Exercises 7.3.1 and 7.3.2).

It should also be pointed out that the representation of linear fractional functions by matrices is not unique. The fraction

$$\frac{as+b}{cs+d} \quad \text{is equal to} \quad \frac{kas+kb}{kcs+kd} \quad \text{for any } k \neq 0.$$

Hence, the function $f(s) = \frac{as+b}{cs+d}$ is represented not only by

$$M = \begin{pmatrix} a & b \\ c & d \end{pmatrix} \quad \text{but also by} \quad kM = \begin{pmatrix} ka & kb \\ kc & kd \end{pmatrix} \quad \text{for any } k \neq 0.$$

The linear transformations kM combine the transformation M with dilation by factor k, so they are all different. Thus, the same linear fractional function f is represented by infinitely many different transformations kM of $\mathbb{R}^2$. The message, again, is that we care only that each of these transformations sends the line with label s to the line with label $f(s)$.

Exercises

7.3.1 Write down a matrix M that represents the map $s \mapsto 1/s$ of $\mathbb{RP}^1$.

7.3.2 Verify that your matrix M in Exercise 7.3.1 maps the y-axis onto the x-axis.

7.3.3 Sketch a picture of the lines with labels $1/2$, $-1/2$, $1/3$, and $-1/3$.

The nonuniqueness of the matrix M corresponding to the linear fractional function f raises the question: Is there a natural way to choose *one* matrix for each linear fractional function? Actually, no, but there is a natural way to choose *two* matrices.

7.3.4 Given that

$$M = \begin{pmatrix} a & b \\ c & d \end{pmatrix} \quad \text{and} \quad ad - bc \neq 0,$$

show that the determinant of kM has absolute value 1 for exactly two of the matrices kM, where $k \neq 0$.

7.4 Spherical geometry

The unit sphere in $\mathbb{R}^3$ consists of all points at unit distance from O, that is, all points (x, y, z) satisfying the equation

$$x^2 + y^2 + z^2 = 1.$$

This surface is also called the *2-sphere*, or $\mathbb{S}^2$, because its points can be described by two coordinates—latitude and longitude, for example. Its geometry is essentially *two-dimensional*, like that of the Euclidean plane $\mathbb{R}^2$ or the real projective plane $\mathbb{RP}^2$, and indeed the fundamental objects of spherical geometry are "points" (ordinary points on the sphere) and "lines" (great circles on the sphere).

However, like the projective plane $\mathbb{RP}^2$, the sphere $\mathbb{S}^2$ is best understood via properties of the three-dimensional space $\mathbb{R}^3$. In particular, the "lines" on $\mathbb{S}^2$ are the intersections of $\mathbb{S}^2$ with planes through O in $\mathbb{R}^3$—the great circles—and the isometries of $\mathbb{S}^2$ are precisely the isometries of $\mathbb{R}^3$ that leave O fixed.

By definition, an isometry f of $\mathbb{R}^3$ preserves distance. Hence, if f leaves O fixed, it sends each point at distance 1 from O to another point at distance 1 from O. In other words, an isometry f of $\mathbb{R}^3$ that fixes O also maps $\mathbb{S}^2$ into itself. The restriction of f to $\mathbb{S}^2$ is therefore an isometry of $\mathbb{S}^2$, because f preserves distances on $\mathbb{S}^2$ as it does everywhere else. This statement is true whether one uses the straight-line distance between points of $\mathbb{S}^2$ or, as is more natural, the great-circle distance along the curved surface of $\mathbb{S}^2$ (Figure 7.4). The isometries of $\mathbb{S}^2$ are the maps of $\mathbb{S}^2$ into itself that preserve great circle distance, and we will see next why they are all restrictions of isometries of $\mathbb{R}^3$.

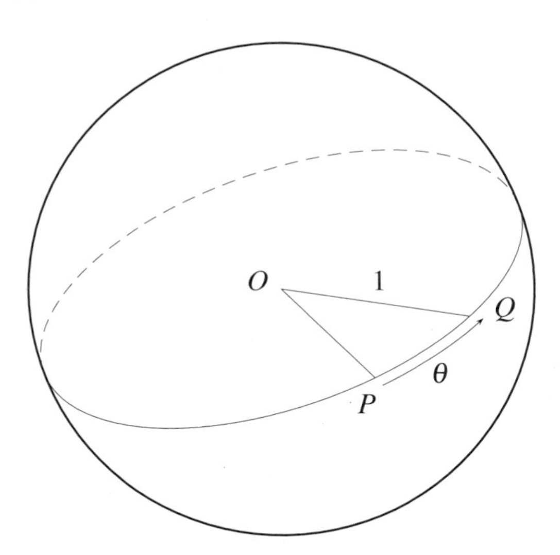

Figure 7.4: Great-circle distance

The isometries of $\mathbb{S}^2$

The simplest isometries of $\mathbb{R}^3$ that fix O are reflections in planes through O. The corresponding isometries of $\mathbb{S}^2$ are the *reflections in great circles*. Two planes $\mathscr{P}_1$ and $\mathscr{P}_2$ meet in a line $\mathscr{L}$ through O, and the product of reflections in $\mathscr{P}_1$ and $\mathscr{P}_2$ is a rotation about $\mathscr{L}$ (through twice the angle between $\mathscr{P}_1$ and $\mathscr{P}_2$). This situation is completely analogous to that in $\mathbb{R}^2$, where the product of reflections through O is a rotation (through twice the angle between the lines).

Finally, there are products of reflections in three planes that are different from products of reflections in one or two planes. One such isometry is the *antipodal map* sending each point (x,y,z) to its *antipodal point* $(-x,-y,-z)$. This map is the product of

- reflection in the (y,z)-plane, which sends (x,y,z) to $(-x,y,z)$,
- reflection in the (z,x)-plane, which sends (x,y,z) to $(x,-y,z)$,
- reflection in the (x,y)-plane, which sends (x,y,z) to $(x,y,-z)$.

As in $\mathbb{R}^2$, there is a "three reflections theorem" that any isometry of $\mathbb{S}^2$ is the product of one, two, or three reflections. The proof is similar to the proof for $\mathbb{R}^2$ in Sections 3.3 and 3.7 (see the exercises below). This three reflections theorem shows why all isometries of $\mathbb{S}^2$ are restrictions of isometries of $\mathbb{R}^3$, namely, because this is true of reflections in great circles.

Exercises

The proof of the three reflections theorem begins, as it did for $\mathbb{R}^2$, by considering the equidistant set of two points.

7.4.1 Show that the equidistant set of two points in $\mathbb{R}^3$ is a plane. Show also that the plane passes through O if the two points are both at distance 1 from O.

7.4.2 Deduce from Exercise 7.4.1 that the equidistant set of two points on $\mathbb{S}^2$ is a "line" (great circle) on $\mathbb{S}^2$.

Next, we establish that there is a unique point on $\mathbb{S}^2$ at given distances from three points not in a "line."

7.4.3 Suppose that two points $P,Q \in \mathbb{S}^2$ have the same distances from three points $A,B,C \in \mathbb{S}^2$ not in a "line." Deduce from Exercise 7.4.2 that $P=Q$.

7.4.4 Deduce from Exercise 7.4.3 that an isometry of $\mathbb{S}^2$ is determined by the images of three points A,B,C not in a "line."

Thus, it remains to show the following. Any three points $A,B,C \in \mathbb{S}^2$ not in a "line" can be mapped to any other three points $A',B',C' \in \mathbb{S}^2$, which are separated by the same respective distances, by one, two, or three reflections.

7.4.5 Complete this proof of the three reflections theorem by imitating the argument in Section 3.7.

7.5 The rotation group of the sphere

The group $\mathrm{Isom}(\mathbb{S}^2)$ of all isometries of $\mathbb{S}^2$ has a subgroup $\mathrm{Isom}^+(\mathbb{S}^2)$ consisting of the isometries that are products of an even number of reflections. Like $\mathrm{Isom}^+(\mathbb{R}^2)$, this is the "orientation-preserving" subgroup. But, unlike $\mathrm{Isom}^+(\mathbb{R}^2)$, $\mathrm{Isom}^+(\mathbb{S}^2)$ includes no "translations"—only rotations. We already know that the product of two reflections of $\mathbb{S}^2$ is a rotation. Hence, to show that the product of any even number of reflections is a rotation, it remains to show that *the product of any two rotations of* $\mathbb{S}^2$ *is a rotation.*

Suppose that the two rotations of $\mathbb{S}^2$ are

- a rotation through angle θ about point P (that is, a rotation with axis through P and its antipodal point $-P$),
- a rotation through angle φ about point Q.

We have established that a rotation through θ about P is the product of reflections in "lines" (great circles) through P. Moreover, they can be *any* "lines" $\mathscr{L}$ and $\mathscr{M}$ through P as long as the angle between $\mathscr{L}$ and $\mathscr{M}$ is $\theta/2$. In particular, we can take the line $\mathscr{M}$ to go through P and Q. Similarly, a rotation through φ about Q is the product of reflections in *any* lines through Q meeting at angle $\varphi/2$, so we can take the first "line" to be $\mathscr{M}$. The second "line" of reflection through Q is then the "line" $\mathscr{N}$ at angle $\varphi/2$ from $\mathscr{M}$ (Figure 7.5).

If $\overline{r}_{\mathscr{L}}, \overline{r}_{\mathscr{M}}, \overline{r}_{\mathscr{N}}$ denote the reflections in $\mathscr{L}, \mathscr{M}, \mathscr{N}$, respectively, then

$$\text{rotation through } \theta \text{ about } P = \overline{r}_{\mathscr{M}}\overline{r}_{\mathscr{L}},$$
$$\text{rotation through } \varphi \text{ about } Q = \overline{r}_{\mathscr{N}}\overline{r}_{\mathscr{M}}.$$

(Bear in mind that products of transformations are read from right to left, as this is the order in which functions are applied.) Hence, the product of these rotations is

$$\overline{r}_{\mathscr{N}}\overline{r}_{\mathscr{M}}\overline{r}_{\mathscr{M}}\overline{r}_{\mathscr{L}} = \overline{r}_{\mathscr{N}}\overline{r}_{\mathscr{L}}, \quad \text{because } \overline{r}_{\mathscr{M}}\overline{r}_{\mathscr{M}} \text{ is the identity.}$$

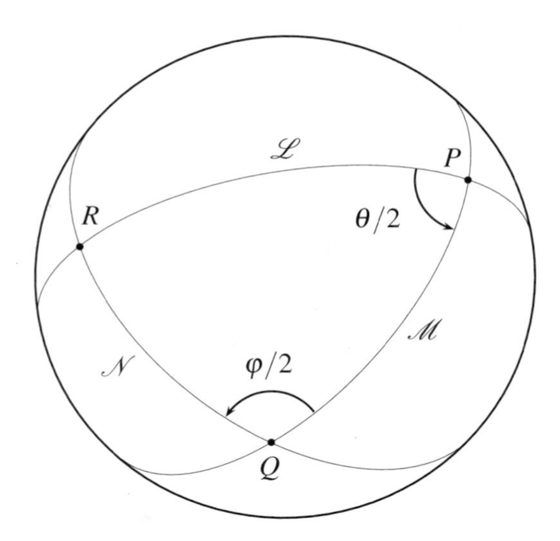

Figure 7.5: Reflection “lines” on the sphere

And it is clear from Figure 7.5 that $\overline{r}_{\mathscr{N}}\overline{r}_{\mathscr{L}}$ is a rotation (about the point R where $\mathscr{L}$ meets $\mathscr{N}$). □

Some special rotations

Before trying to obtain an overview of the rotation group of the sphere, it is helpful to look at the rotation group of the circle, which is analogous but considerably simpler.

The circle can be viewed as the unit *one-dimensional sphere* $\mathbb{S}^1$ in $\mathbb{R}^2$, and its rotations are products of reflections in lines through O. This circumstance is what makes the rotation group of the circle similar to the rotation group of the sphere. What makes it a lot simpler is the fact that *each rotation of* $\mathbb{S}^1$ *corresponds to a point of* $\mathbb{S}^1$, because each rotation of $\mathbb{S}^1$ is determined by the point to which it sends the specific point $(1,0)$. In other words, each rotation of the circle corresponds to an *angle*, namely the angle between the initial and final positions of any line through O. Also, rotations of $\mathbb{S}^1$ *commute*, because rotation through θ followed by rotation through φ results in rotation through $\theta+\varphi$, which is also the result of rotation through φ followed by rotation through θ.

In contrast to $\mathbb{S}^1$, a rotation of $\mathbb{S}^2$ depends on three numbers: two angles that give the direction of its *axis*, and the amount of turn about this axis. Thus, the rotations of $\mathbb{S}^2$ cannot correspond to the points of $\mathbb{S}^2$, although they do correspond to the points of an interesting three-dimensional space, as we shall see in Section 7.6.

Rotations of $\mathbb{S}^2$ generally do *not* commute, as can be seen by combining a quarter turn $z_{1/4}$ around the z-axis with a half-turn $x_{1/2}$ around the x-axis. Supposing that the quarter turn is in the direction that takes $(1,0,0)$ to $(0,1,0)$, we have

$$(1,0,0) \overset{z_{1/4}}{\longmapsto} (0,1,0) \overset{x_{1/2}}{\longmapsto} (0,-1,0),$$

whereas

$$(1,0,0) \overset{x_{1/2}}{\longmapsto} (1,0,0) \overset{z_{1/4}}{\longmapsto} (0,1,0).$$

Exercises

In the Euclidean plane $\mathbb{R}^2$, the product of a rotation about a point P and a rotation about a point Q is *not* necessarily a rotation.

7.5.1 Give an example of two rotations of $\mathbb{R}^2$ whose product is a translation.

7.5.2 By imitating the construction of rotations of $\mathbb{S}^2$ via reflections, explain how to decide whether the product of two rotations of $\mathbb{R}^2$ is a rotation and, if so, how to find its center and angle.

The group of all isometries of $\mathbb{R}^2$, unlike the group of rotations of $\mathbb{R}^2$ about O, is not commutative.

7.5.3 Find a rotation and reflection of $\mathbb{R}^2$ that do not commute.

7.6 Representing space rotations by quaternions

The most elegant (and practical) way to describe rotations of $\mathbb{R}^3$ or $\mathbb{S}^2$ is with the help of the quaternions, which were introduced in Section 6.6. Because they appeared there only in exercises, we now review their basic properties for the sake of completeness.

A *quaternion* is a 2×2 matrix of the form

$$\mathbf{q} = \begin{pmatrix} a+ib & c+id \\ -c+id & a-ib \end{pmatrix}, \quad \text{where } a,b,c,d \in \mathbb{R} \text{ and } i^2 = -1.$$

We also write $\mathbf{q}$ in the form $\mathbf{q} = a\mathbf{1} + b\mathbf{i} + c\mathbf{j} + d\mathbf{k}$, where

$$\mathbf{1} = \begin{pmatrix} 1 & 0 \\ 0 & 1 \end{pmatrix}, \quad \mathbf{i} = \begin{pmatrix} i & 0 \\ 0 & -i \end{pmatrix}, \quad \mathbf{j} = \begin{pmatrix} 0 & 1 \\ -1 & 0 \end{pmatrix}, \quad \mathbf{k} = \begin{pmatrix} 0 & i \\ i & 0 \end{pmatrix}.$$

The various products of $\mathbf{i}$, $\mathbf{j}$, and $\mathbf{k}$ are easily worked out by matrix multiplication, and one finds for example that $\mathbf{ij} = \mathbf{k} = -\mathbf{ji}$ and $\mathbf{i}^2 = -\mathbf{1}$.

Because $\mathbf{q}$ corresponds to the quadruple (a,b,c,d) of real numbers, we can view $\mathbf{q}$ as a point in $\mathbb{R}^4$. If $\mathbf{p}$ is an arbitrary point in $\mathbb{R}^4$ then the map sending $\mathbf{p} \mapsto \mathbf{pq}$ multiplies all distances in $\mathbb{R}^4$ by $|\mathbf{q}|$, the distance of $\mathbf{q}$ from the origin. To see why, notice that

$$\det \mathbf{q} = a^2 + b^2 + c^2 + d^2 = |\mathbf{q}|^{\mathbf{2}}.$$

Then it follows from the multiplicative property of determinants that

$$|\mathbf{pq}|^{\mathbf{2}} = \det(\mathbf{pq}) = (\det \mathbf{p})(\det \mathbf{q}) = |\mathbf{p}|^{\mathbf{2}}|\mathbf{q}|^{\mathbf{2}} \quad \text{and hence} \quad |\mathbf{pq}| = |\mathbf{p}||\mathbf{q}|.$$

It follows that, for any points $\mathbf{p}_1, \mathbf{p}_2 \in \mathbb{R}^4$,

$$|\mathbf{p}_1\mathbf{q} - \mathbf{p}_2\mathbf{q}| = |(\mathbf{p}_1 - \mathbf{p}_2)\mathbf{q}| = |\mathbf{p}_1 - \mathbf{p}_2||\mathbf{q}|.$$

Hence, the distance $|\mathbf{p}_1 - \mathbf{p}_2|$ between any two points is multiplied by the constant $|\mathbf{q}|$. In particular, *if* $|\mathbf{q}| = \mathbf{1}$, *then the map* $\mathbf{p} \mapsto \mathbf{pq}$ *is an isometry of* $\mathbb{R}^4$.

The map $\mathbf{p} \mapsto \mathbf{qp}$ (which is not necessarily the same as the map $\mathbf{p} \mapsto \mathbf{pq}$, because quaternion multiplication is not commutative) is likewise an isometry when $|\mathbf{q}| = 1$. These maps are useful for studying rotations of $\mathbb{R}^4$ but, more surprisingly, also for studying rotations of $\mathbb{R}^3$.

Rotations of $(\mathbf{i}, \mathbf{j}, \mathbf{k})$-space

If $\mathbf{p}$ is any quaternion in $(\mathbf{i}, \mathbf{j}, \mathbf{k})$-space,

$$\mathbf{p} = x\mathbf{i} + y\mathbf{j} + z\mathbf{k}, \quad \text{where } x, y, z \in \mathbb{R},$$

and if $\mathbf{q}$ is any nonzero quaternion, then it turns out that $\mathbf{qpq}^{-1}$ also lies in $(\mathbf{i}, \mathbf{j}, \mathbf{k})$-space. Thus, *if* $|\mathbf{q}| = 1$, *then the map* $\mathbf{p} \mapsto \mathbf{qpq}^{-1}$ *defines an isometry of* $\mathbb{R}^3$, because $(\mathbf{i}, \mathbf{j}, \mathbf{k})$-space is just the space of real triples (x, y, z) and hence a copy of $\mathbb{R}^3$.

Moreover, any quaternion with $|\mathbf{q}| = 1$ can be written in the form

$$\mathbf{q} = \cos\frac{\theta}{2} + (l\mathbf{i} + m\mathbf{j} + n\mathbf{k})\sin\frac{\theta}{2}, \quad \text{where} \quad l^2 + m^2 + n^2 = 1,$$

and *the isometry* $\mathbf{p} \mapsto \mathbf{q}\mathbf{p}\mathbf{q}^{-1}$ *is a rotation of* $(\mathbf{i},\mathbf{j},\mathbf{k})$*-space through angle* θ *about the axis through* $\mathbf{0}$ *and* $l\mathbf{i} + m\mathbf{j} + n\mathbf{k}$.

These facts can be confirmed by calculation, but we verify them only for the special case in which the axis of rotation is in the $\mathbf{i}$ direction, and for special points $\mathbf{p}$ that easily determine the nature of the isometry. Notice how the angles $\theta/2$ in $\mathbf{q}$ and $\mathbf{q}^{-1}$ combine to produce angle of rotation θ.

Example. The map $\mathbf{p} \mapsto \mathbf{q}\mathbf{p}\mathbf{q}^{-1}$, where $\mathbf{q} = \cos\frac{\theta}{2} + \mathbf{i}\sin\frac{\theta}{2}$.

First we check that any point $x\mathbf{i}$ on the $\mathbf{i}$-axis is fixed by this map.

$$\begin{aligned}
\mathbf{q}x\mathbf{i}\mathbf{q}^{-1} &= \left(\cos\frac{\theta}{2} + \mathbf{i}\sin\frac{\theta}{2}\right) x\mathbf{i} \left(\cos\frac{\theta}{2} - \mathbf{i}\sin\frac{\theta}{2}\right) \\
&= \left(\cos\frac{\theta}{2} + \mathbf{i}\sin\frac{\theta}{2}\right)\left(x\mathbf{i}\cos\frac{\theta}{2} + x\mathbf{1}\sin\frac{\theta}{2}\right) \qquad \text{because } \mathbf{i}^2 = -\mathbf{1} \\
&= x\mathbf{i}\left(\cos^2\frac{\theta}{2} + \sin^2\frac{\theta}{2}\right) + \mathbf{1}\left(\sin\frac{\theta}{2}\cos\frac{\theta}{2} - \sin\frac{\theta}{2}\cos\frac{\theta}{2}\right) \\
&= x\mathbf{i}.
\end{aligned}$$

Next we check that the point $\mathbf{j}$ is rotated through angle θ in the $(\mathbf{j},\mathbf{k})$-plane, to the point $\mathbf{j}\cos\theta + \mathbf{k}\sin\theta$.

$$\begin{aligned}
\mathbf{q}\mathbf{j}\mathbf{q}^{-1} &= \left(\cos\frac{\theta}{2} + \mathbf{i}\sin\frac{\theta}{2}\right)\mathbf{j}\left(\cos\frac{\theta}{2} - \mathbf{i}\sin\frac{\theta}{2}\right) \\
&= \left(\cos\frac{\theta}{2} + \mathbf{i}\sin\frac{\theta}{2}\right)\left(\mathbf{j}\cos\frac{\theta}{2} + \mathbf{k}\sin\frac{\theta}{2}\right) \qquad \text{because } \mathbf{j}\mathbf{i} = -\mathbf{k} \\
&= \mathbf{j}\left(\cos^2\frac{\theta}{2} - \sin^2\frac{\theta}{2}\right) + \mathbf{k}\left(2\sin\frac{\theta}{2}\cos\frac{\theta}{2}\right) \qquad \text{because } \mathbf{i}\mathbf{k} = \mathbf{j},\ \mathbf{i}\mathbf{j} = \mathbf{k} \\
&= \mathbf{j}\cos\theta + \mathbf{k}\sin\theta.
\end{aligned}$$

It can be similarly checked that $\mathbf{q}\mathbf{k}\mathbf{q}^{-1} = -\mathbf{k}\sin\theta + \mathbf{j}\cos\theta$. Hence the isometry $\mathbf{p} \mapsto \mathbf{q}\mathbf{p}\mathbf{q}^{-1}$ is a rotation of the $(\mathbf{j},\mathbf{k})$-plane through θ, because this is certainly what such a rotation does to the points $\mathbf{0}$, $\mathbf{j}$, and $\mathbf{k}$, and we know from Section 3.7 that any isometry of a plane is determined by what it does to three points not in a line.

Thus, the isometry $\mathbf{p} \mapsto \mathbf{q}\mathbf{p}\mathbf{q}^{-1}$ of $(\mathbf{i}, \mathbf{j}, \mathbf{k})$-space leaves the $\mathbf{i}$-axis fixed and rotates the $(\mathbf{j}, \mathbf{k})$-plane through angle θ, so it is a rotation through θ about the $\mathbf{i}$-axis. □

It should be emphasized that if the quaternion $\mathbf{q}$ represents a certain rotation of $\mathbb{R}^3$, then so does the opposite quaternion $-\mathbf{q}$, because $\mathbf{q}\mathbf{p}\mathbf{q}^{-1} = (-\mathbf{q})\mathbf{p}(-\mathbf{q})^{-1}$. Thus, rotations of $\mathbb{R}^3$ actually correspond to *pairs* of quaternions $\pm\mathbf{q}$ with $|\mathbf{q}| = 1$. This has interesting consequences when we try to interpret the group of rotations of $\mathbb{R}^3$ as a geometric object in its own right (Section 7.8).

Exercises

The representation of space rotations by quaternions is analogous to the representation of plane rotations by complex numbers, which was described in Section 4.7. As a warmup for the study of a *finite group of space rotations* in Section 7.7, we look here at some finite groups of plane rotations and the geometric objects they preserve. We take the plane to be $\mathbb{C}$, the complex numbers.

7.6.1 Consider the square with vertices 1, i, -1, and $-i$. There is a group of four rotations of $\mathbb{C}$ that map the square onto itself. These rotations correspond to multiplying $\mathbb{C}$ by which four numbers?

7.6.2 The *cyclic group* C_n is the group of n rotations that maps a regular n-gon onto itself. These rotations correspond to multiplying $\mathbb{C}$ by which n complex numbers?

The noncommutative multiplication of quaternions is a blessing when we want to use them to represent space rotations, because we know that products of space rotations do not generally commute. Nevertheless, one wonders whether there is a reasonable commutative "product" operation on any $\mathbb{R}^n$, for any $n \geq 3$. "Reasonable" here includes the property $|uv| = |u||v|$ that holds for products on $\mathbb{R}$ and $\mathbb{R}^2$ (the real and complex numbers), and the field axioms from Section 6.5.

In particular, there should be a *multiplicative identity*: a point $\mathbf{1}$ such that $|\mathbf{1}| = 1$ and $\mathbf{u}\mathbf{1} = \mathbf{u}$ for any point $\mathbf{u}$. Moreover, because $n \geq 3$, we can find points $\mathbf{i}$ and $\mathbf{j}$, also of absolute value 1, such that $\mathbf{1}$, $\mathbf{i}$, and $\mathbf{j}$ are in mutually perpendicular directions from O. Figure 7.6 shows these points, together with their negatives.

7.6.3 Show that $|\mathbf{1}+\mathbf{i}| = \sqrt{2} = |\mathbf{1}-\mathbf{i}|$, and deduce from the assumptions about the product operation that $2 = |\mathbf{1}-\mathbf{i}^2|$, which means that the point $\mathbf{1}-\mathbf{i}^2$ is at distance 2 from O.

7.6.4 Show also that $|\mathbf{i}^2| = 1$, so the point $\mathbf{1}-\mathbf{i}^2$ is at distance 1 from $\mathbf{1}$. Conclude from this and Exercise 7.6.3 that $\mathbf{i}^2 = -\mathbf{1}$.

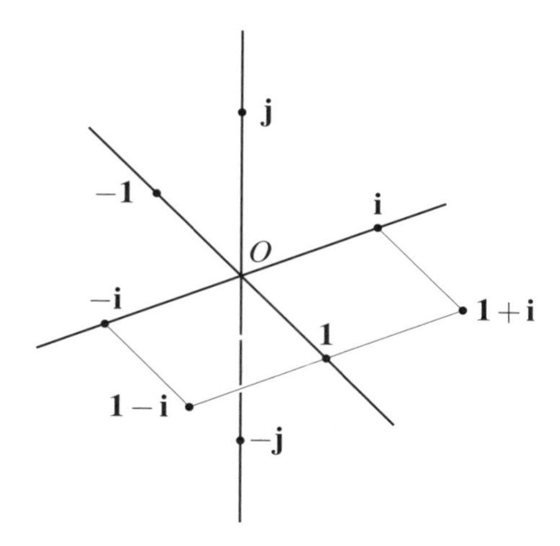

Figure 7.6: Points in perpendicular directions from O

7.6.5 Show similarly that $\mathbf{u}^2 = -\mathbf{1}$ for any point $\mathbf{u}$ whose direction from O is perpendicular to the direction of $\mathbf{1}$, and whose absolute value is 1.

7.6.6 Given that $\mathbf{i}$ and $\mathbf{j}$ are in perpendicular directions, show (multiplying the whole space by $\mathbf{i}$) that so are $\mathbf{i}^2 = -\mathbf{1}$ and $\mathbf{ij}$, and hence so too are $\mathbf{1}$ and $\mathbf{ij}$.

7.6.7 Thus, $\mathbf{ij}$ is one point $\mathbf{u}$ for which $\mathbf{u}^2 = -\mathbf{1}$, by Exercise 7.6.5. Deduce that

$$-\mathbf{1} = (\mathbf{ij})^2 = (\mathbf{ij})(\mathbf{ij}) = \mathbf{jiij} \quad \text{by the commutative and associative laws}$$

and show that this leads to the contradiction $-\mathbf{1} = \mathbf{1}$.

Therefore, when $n \geq 3$, there is no product on $\mathbb{R}^n$ that satisfies all the field axioms.

7.7 A finite group of space rotations

$\mathbb{R}^3$ is home to the *regular polyhedra*, the remarkable symmetric objects discussed in Section 1.6 and shown in Figure 1.19. The best known of them is the cube, and the simplest of them is the *tetrahedron*, which fits inside the cube as shown in Figure 7.7.

Also shown in this figure are some rotations of the tetrahedron that are called *symmetries* because they preserve its appearance. If we choose a fixed position of the tetrahedron—a "tetrahedral hole" in space as it were—then these rotations bring the tetrahedron to positions where it once again

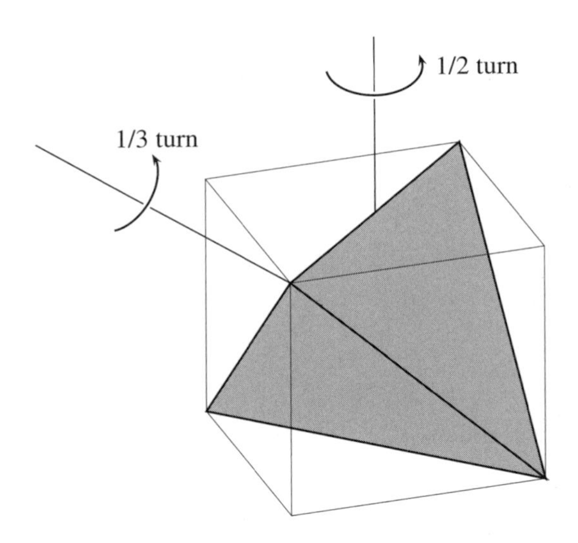

Figure 7.7: The tetrahedron and axes of rotation

fits in the hole. Altogether there are 12 such rotations. We can choose any one of the four faces to match a fixed face of the hole, say, the *front* face. Each of the four faces that can go in front has three edges that can match a given edge, say, the *bottom* edge, in the front face of the hole. Thus, we have $4 \times 3 = 12$ ways in which the tetrahedron can occupy the same position, each corresponding to a different symmetry. But once we have chosen a particular face to go in front, and a particular edge of that face to go on the bottom, we know where everything goes, so the symmetry is completely determined. Hence, there are exactly 12 rotational symmetries.

Each symmetry can be obtained, from a given initial position, by rotations like those shown in Figure 7.7. First there is the *trivial rotation*, which gives the *identity symmetry*, obtained by rotation through angle zero (about any axis). Then there are 11 nontrivial rotations, divided into two different types:

- The first type is a 1/2 turn about an axis through centers of opposite edges of the tetrahedron (which also goes through opposite face centers of the cube). There are three such axes. Hence, there are three rotations of this type.

- The second type is a 1/3 turn about an axis through a vertex and the center of the face opposite to it (which also goes through opposite vertices of the cube). There are four such axes, and hence eight rotations of this type—because the 1/3 turn clockwise is different from the 1/3 turn anticlockwise.

Notice also that each 1/2 turn moves all four vertices, whereas each 1/3 turn leaves one vertex fixed and moves the remaining three. Thus, the 11 nontrivial rotations are all different. Therefore, together with the trivial rotation, they account for all 12 symmetries of the tetrahedron.

The quaternions representing rotations of the tetrahedron

As explained in Section 7.6, a rotation of $(\mathbf{i},\mathbf{j},\mathbf{k})$-space through angle θ about axis $l\mathbf{i}+m\mathbf{j}+n\mathbf{k}$ corresponds to a quaternion pair $\pm\mathbf{q}$, where

$$\mathbf{q} = \cos\frac{\theta}{2} + (l\mathbf{i}+m\mathbf{j}+n\mathbf{k})\sin\frac{\theta}{2}.$$

If we choose coordinate axes so that the sides of the cube in Figure 7.7 are parallel to the $\mathbf{i}$, $\mathbf{j}$, and $\mathbf{k}$ axes, then the axes of rotation are virtually immediate, and the corresponding quaternions are easy to work out.

- We can take the lines through opposite face centers of the cube to be the $\mathbf{i}$, $\mathbf{j}$, and $\mathbf{k}$ axes. For a 1/2 turn, the angle $\theta = \pi$, and hence $\theta/2 = \pi/2$. Therefore, because $\cos\frac{\pi}{2} = 0$ and $\sin\frac{\pi}{2} = 1$, the 1/2 turns about the $\mathbf{i}$, $\mathbf{j}$, and $\mathbf{k}$ axes are given by the quaternions $\mathbf{i}$, $\mathbf{j}$, and $\mathbf{k}$ themselves.

  Thus, the three 1/2 turns are represented by the three quaternion pairs

$$\pm\mathbf{i}, \quad \pm\mathbf{j}, \quad \pm\mathbf{k}.$$

- Given the choice of $\mathbf{i}$, $\mathbf{j}$, and $\mathbf{k}$ axes, the four rotation axes through opposite vertices of the cube correspond to four quaternion pairs, which together make up the eight combinations

$$\frac{1}{\sqrt{3}}(\pm\mathbf{i}\pm\mathbf{j}\pm\mathbf{k}) \quad \text{(independent choices of + or − sign).}$$

  The factor $\frac{1}{\sqrt{3}}$ is to give each of these quaternions the absolute value 1, as specified for the representation of rotations.

For each 1/3 turn, we have $\theta = \pm 2\pi/3$. Hence,

$$\cos\frac{\theta}{2} = \cos\frac{\pi}{3} = \frac{1}{2}, \qquad \sin\frac{\theta}{2} = \pm\sin\frac{\pi}{3} = \pm\frac{\sqrt{3}}{2}.$$

The $\sqrt{3}$ in $\sin\frac{\pi}{3}$ neatly cancels the factor $1/\sqrt{3}$ in the axis of rotation, and we find that the eight 1/3 turns are represented by the eight pairs of opposites among the 16 quaternions

$$\pm\frac{1}{2} \pm \frac{\mathbf{i}}{2} \pm \frac{\mathbf{j}}{2} \pm \frac{\mathbf{k}}{2}.$$

Finally, the identity rotation is represented by the pair ± 1, and thus the 12 symmetries of the tetrahedron are represented by the 24 quaternions

$$\pm 1, \quad \pm\mathbf{i}, \quad \pm\mathbf{j}, \quad \pm\mathbf{k}, \quad \pm\frac{1}{2} \pm \frac{\mathbf{i}}{2} \pm \frac{\mathbf{j}}{2} \pm \frac{\mathbf{k}}{2}.$$

The 24-cell

These 24 quaternions all lie at distance 1 from O in $\mathbb{R}^4$, and they are distributed in a highly symmetrical manner. In fact, they are the vertices of a four-dimensional figure analogous to a regular polyhedron—called a *regular polytope*. This particular polytope is called the *24-cell*. Because we cannot directly perceive four-dimensional figures, the best we can do is study the 24-cell via projections of it into $\mathbb{R}^3$ (just as we often study polyhedra, such as the tetrahedron and cube, via projections onto the plane such as Figure 7.7). One such projection is shown in Figure 7.8 (which of course is a projection of a three-dimensional figure onto the plane—but it is easy to visualize what the three-dimensional figure is). This superb drawing is taken from Hilbert and Cohn-Vossen's *Geometry and the Imagination*.

Exercises

The vertices of the 24-cell include the eight unit points (positive and negative) on the four axes in $\mathbb{R}^4$, but the other 16 points have some less obvious properties.

7.7.1 Verify directly that the 16 points $\pm\frac{1}{2} \pm \frac{\mathbf{i}}{2} \pm \frac{\mathbf{j}}{2} \pm \frac{\mathbf{k}}{2}$ are all at distance 1 from the origin in $\mathbb{R}^4$.

7.7.2 Deduce that the distance from the center to any vertex of a four-dimensional cube is equal to the length of its side.

7.7.3 Show also that each of the points $\pm\frac{1}{2} \pm \frac{\mathbf{i}}{2} \pm \frac{\mathbf{j}}{2} \pm \frac{\mathbf{k}}{2}$ is at distance 1 from the four nearest unit points on the axes.

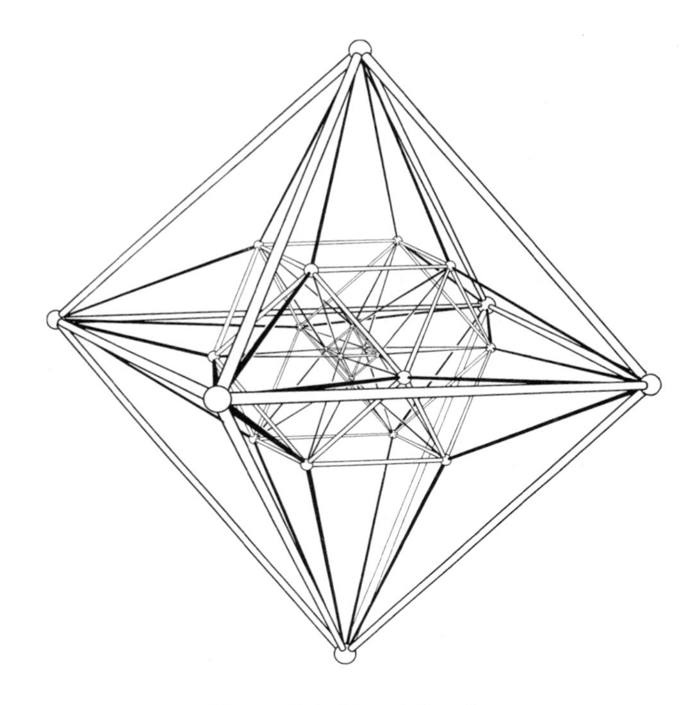

Figure 7.8: The 24-cell

7.8 The groups $\mathbb{S}^3$ and $\mathbb{RP}^3$

The rotations of the tetrahedron, which were discussed in Section 7.7, vividly show that *a group of rotations is itself a geometric object*. This statement is just as true of the group of all rotations of $\mathbb{S}^2$. In fact, this group is closely related to two important geometric objects: the *3-sphere* $\mathbb{S}^3$ and the *three-dimensional real projective space* $\mathbb{RP}^3$.

Just as the 1- and 2-spheres are the sets of points at unit distance from O in $\mathbb{R}^2$ and $\mathbb{R}^3$ respectively, the 3-sphere is the set of points in $\mathbb{R}^4$ at unit distance from O:

$$\mathbb{S}^3 = \{(a,b,c,d) \in \mathbb{R}^4 : a^2+b^2+c^2+d^2 = 1\}.$$

The points (a,b,c,d) on $\mathbb{S}^3$ correspond to quaternions $\mathbf{q} = a\mathbf{1}+b\mathbf{i}+c\mathbf{j}+d\mathbf{k}$ with $|\mathbf{q}| = 1$, because $|\mathbf{q}|^2 = a^2+b^2+c^2+d^2$. Hence, *rotations of $\mathbb{S}^2$, which correspond to pairs $\pm\mathbf{q}$ of such quaternions, correspond to point pairs $\pm(a,b,c,d)$ on $\mathbb{S}^3$.*

And to what else do the point pairs $\pm(a,b,c,d)$ correspond? Well, remember from Section 5.4 how we described the real projective space $\mathbb{RP}^3$. Its "points" are lines through O in $\mathbb{R}^4$. But a line through O in $\mathbb{R}^4$ meets $\mathbb{S}^3$ in a pair of antipodal points $\pm(a,b,c,d)$. Thus, *it is also valid to view the point pairs* $\pm(a,b,c,d)$ *on* $\mathbb{S}^3$ *as single "points" of* $\mathbb{RP}^3$. Hence, rotations of $\mathbb{S}^2$ correspond to points of $\mathbb{RP}^3$, and the group of all rotations of $\mathbb{S}^2$ is in some sense the "same" as the geometric object $\mathbb{RP}^3$.

To explain what we mean by "sameness" here, we have to say something about groups in general, what it means for two groups to be the "same", and what it means for a geometric object to acquire the structure of a group.

Abstract groups and isomorphisms

We began this chapter with the idea of a *group of transformations*: a collection G of functions on a space S with the properties that

- if $f,g \in G$, then $fg \in G$,
- if $f \in G$, then $f^{-1} \in G$.

The "product" fg of f and g here is the *composite function*, which is defined by $fg(x) = f(g(x))$. However, we have found it convenient to *represent* certain functions, such as rotations, by algebraic objects, such as matrices, whose "product" is defined algebraically.

It is therefore desirable to have a more general concept of group, which does not presuppose that the product operation is function composition. We define an *abstract group* to be a set G, which contains a special element 1 and for each g an element g^{-1}, with a "product" operation satisfying the following axioms:

$$g_1(g_2g_3) = (g_1g_2)g_3 \quad \text{(associativity)}$$
$$g1 = g \quad \text{(identity)}$$
$$gg^{-1} = 1 \quad \text{(inverse)}$$

The associative axiom is automatically satisfied for function composition, because if g_1, g_2, g_3 are functions, then $g_1(g_2g_3)$ and $(g_1g_2)g_3$ both mean the same thing, namely, the function $g_1(g_2(g_3(x)))$. It is also satisfied when the group consists of numbers, because the product of numbers is well known to be associative.

In other cases, the easiest way to prove associativity is to show, if possible, that the group operation corresponds to function composition. For example, the matrix product operation is associative because matrices behave like linear transformations under composition. It follows in turn that the quaternion product operation is associative, because quaternions can be viewed as matrices.

When we say that the elements of a certain group G "correspond to" or "behave like" or "can be viewed as" elements of another group G', we have in mind a precise relationship called an *isomorphism* of G onto G'. The word "isomorphism" comes from the Greek for "same form," and it means that *there is a one-to-one correspondence between G and G′ that preserves products*. That is, an isomorphism is a function

$$\varphi : G \to G' \quad \text{such that} \quad \varphi(g_1 g_2) = \varphi(g_1)\varphi(g_2).$$

For example, the group G of rotations of the circle, under composition of isometries, is isomorphic to group G' of complex numbers of the form

$$\cos\theta + i\sin\theta, \quad \text{under multiplication.}$$

If r_θ denotes the rotation through angle θ, then the isomorphism φ is defined by

$$\varphi(r_\theta) = \cos\theta + i\sin\theta.$$

Sometimes there is a natural one-to-one correspondence φ between a group G and a set S. In that case, we can use φ to *transfer the group structure from G to S*. That is, we *define* the product of elements $\varphi(g_1)$ and $\varphi(g_2)$ to be $\varphi(g_1 g_2)$. Here are some examples.

- The complex numbers $\cos\theta + i\sin\theta$ form a group, and they correspond to the points $(\cos\theta, \sin\theta)$ of the unit circle $\mathbb{S}^1$. Therefore, we can define the product of points $(\cos\theta_1, \sin\theta_1)$ and $(\cos\theta_2, \sin\theta_2)$ on $\mathbb{S}^1$ to be the point corresponding to the product of the corresponding complex numbers. This point is $(\cos(\theta_1+\theta_2), \sin(\theta_1+\theta_2))$.

- Likewise, the quaternions $\mathbf{q} = a\mathbf{1} + b\mathbf{i} + c\mathbf{j} + d\mathbf{k}$ with $|\mathbf{q}| = 1$ form a group, and they correspond to the points (a,b,c,d) of the 3-sphere $\mathbb{S}^3$. Hence, we can define the product of points (a_1,b_1,c_1,d_1) and (a_2,b_2,c_2,d_2) corresponding to the quaternions $\mathbf{q}_1$ and $\mathbf{q}_2$, say, to be the point corresponding to the quaternion $\mathbf{q}_1\mathbf{q}_2$. Under this product operation, $\mathbb{S}^3$ *is a group.*

- Finally, pairs of opposite quaternions $\pm\mathbf{q}$ with $|\mathbf{q}| = 1$ form a group under the operation defined by

$$(\pm\mathbf{q}_1)(\pm\mathbf{q}_2) = \pm\mathbf{q}_1\mathbf{q}_2.$$

We know that these pairs are in one-to-one correspondence with the points of $\mathbb{RP}^3$. Hence, we can transfer the group structure of these quaternion pairs (which is also the group structure of the rotations of $\mathbb{S}^2$) to $\mathbb{RP}^3$. Under the transferred operation, $\mathbb{RP}^3$ *is a group*.

The group structures on $\mathbb{S}^1$, $\mathbb{S}^3$, and $\mathbb{RP}^3$ obtained in this way are particularly interesting because they are *continuous*. That is, if g_1' is near to g_1 and g_2' is near to g_2, then $g_1'g_2'$ is near to g_1g_2. It is known that $\mathbb{S}^2$ does *not* have a continuous group structure, and in fact $\mathbb{S}^1$ and $\mathbb{S}^3$ are the *only* spheres with continuous group structures on them.

7.9 Discussion

The word "geometry" comes from the Greek for "earth measurement," and legend has it that the subject grew from the land measurement concerns of farmers whose land was periodically flooded by the river Nile. As recently as the 18th century, one finds carpenters and other artisans listed among the subscribers to geometry books, so there is no doubt that Euclidean geometry is the geometry of down-to-earth measurement. It continues to be a *tactile* subject today, when one talks about "translating," "rotating," and "moving objects rigidly."

The most *visual* branch of geometry is projective geometry, because it is more concerned with how objects *look* than with what they actually are. It is no surprise that projective geometry originated from the concerns of artists, and that many of its practitioners today work in the fields of video games and computer graphics.

Affine geometry occupies a position in the middle. It also originates from an artistic tradition, but from one less radical than that of Renaissance Italy—the classical art of China and Japan. Chinese and Japanese drawings often adopt unusual viewpoints, where one might expect perspective, but they generally preserve parallels. Typically, the picture is a "projection from infinity," which is an affine map. Figure 7.9 shows an example, a woodblock print by the Japanese artist Suzuki Harunobu from around 1760.

Figure 7.9: Harunobu's *Courtesan on veranda*

Notice that all the parallel lines are shown as parallel, with the result that the (obviously rectangular) panels on the screen appear as identical parallelograms. Likewise, the planks on the veranda appear with parallel edges and equal widths, which creates a certain "flatness" because all parts of the picture seem to be the same distance away from us. Speaking mathematically, they are—because the view is what one would see from infinity with infinite magnification. A similar effect occurs in photographs of distant buildings taken with a large amount of zoom.

Affine maps are also popular in engineering drawing, in which the so-called "axonometric projection" is often used to depict an object in three dimensions while retaining correct proportions in a given direction. The affine picture gives a good compromise between a realistic view and an accurate plan. See Figure 7.10, which shows an axonometric projection of a cube.

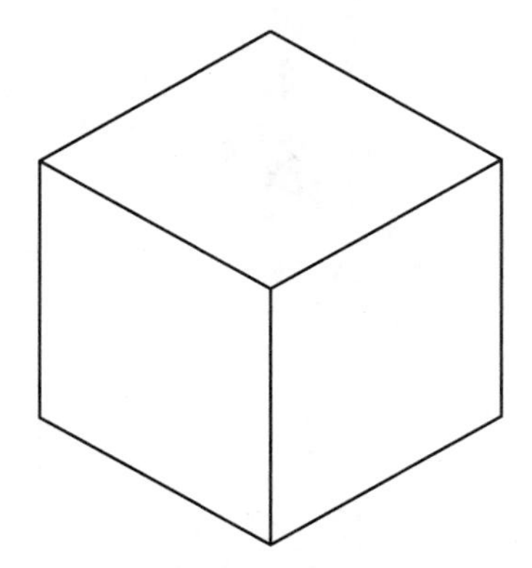

Figure 7.10: Affine view of the cube

The fourth dimension

The discovery of quaternions in 1843 was the first of a series of discoveries that drew attention to spaces of more than three dimensions and to the remarkable properties of $\mathbb{R}^4$ in particular.

From around 1830, the Irish mathematician William Rowan Hamilton had been searching in vain for "n-dimensional number systems" analogous to the real numbers $\mathbb{R}$ and the complex numbers $\mathbb{C}$. Because $\mathbb{C}$ can be viewed as $\mathbb{R}^2$ under vector addition

$$(u_1, u_2) + (v_1, v_2) = (u_1 + v_1, u_2 + v_2)$$

and the multiplication operation

$$(u_1, u_2)(v_1, v_2) = (u_1 v_1 - u_2 v_2, u_1 v_2 + u_2 v_1),$$

Hamilton thought that $\mathbb{R}^3$ could also be viewed as a number system by some clever choice of multiplication rule. He took a "number system" to be what we now call a field, together with an absolute value

$$|\mathbf{u}| = |(u_1, u_2, u_3)| = \sqrt{u_1^2 + u_2^2 + u_3^2}$$

which is multiplicative: $|\mathbf{uv}| = |\mathbf{u}||\mathbf{v}|$.

We now know (for example, by Exercise 7.6.7) that such a system is impossible in any $\mathbb{R}^n$ with $n \geq 3$. But, luckily for Hamilton, it is *almost* possible in $\mathbb{R}^4$. The quaternions satisfy all the field axioms except commutative multiplication, and their absolute value is multiplicative. The only other $\mathbb{R}^n$ that comes close is $\mathbb{R}^8$, where the octonions $\mathbb{O}$ satisfy all the field properties except the commutative and associative laws. (Recall from Section 6.8 that the quaternions and octonions also play an important role in projective geometry.)

Hamilton knew that quaternions give a nice representation of rotations in $\mathbb{R}^3$, but the first to work out the quaternions for the symmetries of regular polyhedra was Cayley in 1863. Cayley's enumeration of these quaternions may be found in his *Mathematical Papers*, volume 5, p. 529. The five regular polyhedra actually exhibit only three types of symmetry—because the cube and octahedron have the same symmetry type, as do the dodecahedron and the icosahedron—which therefore correspond to three highly symmetric sets of quadruples in $\mathbb{R}^4$.

Cayley did not investigate the geometric properties of these point sets in $\mathbb{R}^4$, but in fact they were already known to the Swiss geometer Ludwig Schläfli in 1852. As we have seen, the 12 rotations of the tetrahedron correspond to the 24 vertices of a figure called the 24-cell. It gets this name because it is bounded by 24 identical regular octahedra. It is one of six regular figures in $\mathbb{R}^4$, analogous to the regular polyhedra in $\mathbb{R}^3$, called the *regular polytopes*. They were discovered by Schläfli, who also proved that there are regular figures analogous to the tetrahedron, cube, and octahedron in each $\mathbb{R}^n$, but that $\mathbb{R}^3$ *and* $\mathbb{R}^4$ *are the only* $\mathbb{R}^n$ *containing other regular figures.*

The 24-cell is the simplest of the exceptional regular figures in $\mathbb{R}^4$; the other two are the *120-cell* (bounded by 120 regular dodecahedra) and the *600-cell* (bounded by 600 regular tetrahedra). The 120-cell has 600 vertices, which correspond to the cell centers of the 600-cell, and vice versa, so the two are related "dually" like the dodecahedron and the icosahedron.

Moreover, the 600-cell arises from the icosahedron in the same way that the 24-cell arises from the tetrahedron. Its 120 vertices correspond to 60 pairs of opposite quaternions, each representing a rotational symmetry of the icosahedron. For more on these amazing objects, see my article *The story of the 120-cell*, which can be read online at

`http://www.ams.org/notices/200201/fea-stillwell.pdf`

8

Non-Euclidean geometry

PREVIEW

In previous chapters, we have seen several reasons why there is such a subject as "foundations of geometry." Geometry is fundamentally visual; yet it can be communicated by nonvisual means: by logic, linear algebra, or group theory, for example. The several ways to communicate geometry give several foundations.

But also, *there is more than one geometry.* Section 7.4 gave a hint of this when we briefly discussed the geometry of the sphere in the language of "points" and "lines." It seems reasonable to call great circles "lines" because they are the straightest curves on the sphere; but they certainly do not have all of the properties of Euclid's lines.

This characteristic makes geometry on a sphere a *non-Euclidean* geometry—one that has been known since ancient times. But it was never seen as a challenge to Euclid, probably because the geometry of the sphere is simply a part of three-dimensional Euclidean geometry, where great circles coexist with genuine straight lines.

The real challenge to Euclid emerged from disquiet over the parallel axiom. Many people found it inelegant and wished that it was a consequence of Euclid's other axioms. It is not, because *there is a geometry that satisfies all of Euclid's axioms* except *the parallel axiom.* This is the geometry of a surface called the *non-Euclidean plane.*

The non-Euclidean plane is *not* an artificial construct built only to show that the parallel axiom cannot be proved. It arises in many places, and today one can hardly discuss differential geometry, the theory of complex numbers, and projective geometry without it. In this chapter, we will see how it arises from the real projective line.

8.1 Extending the projective line to a plane

In this book, we have been concerned mainly with the geometry of lines, partly because lines are the foundation of geometry and partly because lines are remarkably interesting. From the regular polygons to the Pappus and Desargues configurations, figures built from lines reveal beautiful connections between geometry and other parts of mathematics. We have seen some of these connections but have barely begun to explore them in depth.

In fact we have not yet gone far toward understanding the geometry of even *one* line—the real projective line $\mathbb{RP}^1$. In Chapter 5, we arrived at an algebraic summary of $\mathbb{RP}^1$ by representing its transformations as linear fractional functions

$$f(x) = \frac{ax+b}{cx+d}, \quad \text{where} \quad a,b,c,d \in \mathbb{R} \quad \text{and} \quad ad-bc \neq 0, \qquad (*)$$

and by uncovering the *cross-ratio*, a "ratio of ratios" left invariant by all linear fractional transformations. But this summary is not as geometric as one would like. It is hard (although perhaps not impossible) to "see" the cross-ratio, and indeed it is hard to see geometric phenomena on the line at all. If only we could extend the projective line in another dimension so that we could see it as a plane! Amazingly, this is possible, and the present chapter shows how.

The idea is to let $\mathbb{RP}^1$ be the boundary ("at infinity") of a plane whose transformations extend the linear fractional transformations of $\mathbb{RP}^1$ in a natural way. Algebra suggests how this should be done. It suggests replacing the real variable x in the linear fractional transformations (*) by a complex variable z and interpreting

$$f(z) = \frac{az+b}{cz+d}$$

as a transformation of the plane $\mathbb{C}$ of complex numbers.

This idea needs a little modification. We should really use transformations of the *upper half plane* of complex numbers $z = x+iy$ with $y > 0$, because the line of real numbers divides $\mathbb{C}$ into two halves. Either half can be taken as the "plane" bounded by the real line, but, when we want to transform one particular half, the extension from x to z is not always the obvious one. For example, the transformation $x \mapsto -x$ of the line should *not* be extended to the transformation $z \mapsto -z$, because the latter maps the upper half plane onto the lower. The correct extension is described in Section 8.2.

Nevertheless, the extension from real to complex numbers works amazingly well. The extended transformations leave invariant a geometric quantity that is clearly visible, namely *angle*. The cross-ratio is also invariant (well, almost), for much the same algebraic reasons as before. And it gives us an invariant *length* in the half plane—something we certainly do not have in the projective line $\mathbb{RP}^1$.

The concept of length that emerges in this way is a little subtle, and it is not as easily visible as angle. It is a *non-Euclidean* measure of length, and it gives rise to *non-Euclidean lines*, which turn out to be the ordinary lines of the form $x =$ constant and the semicircles with their centers on the x-axis. These "lines" are the curves of shortest non-Euclidean length between given points, and they have all the properties of "lines" in Euclid's geometry *except the parallel property*. That is, if $\mathscr{L}$ is any non-Euclidean line and P is a point of the upper half plane outside $\mathscr{L}$ then there is more than one non-Euclidean line through P that does not meet $\mathscr{L}$. Figure 8.1 shows an example.

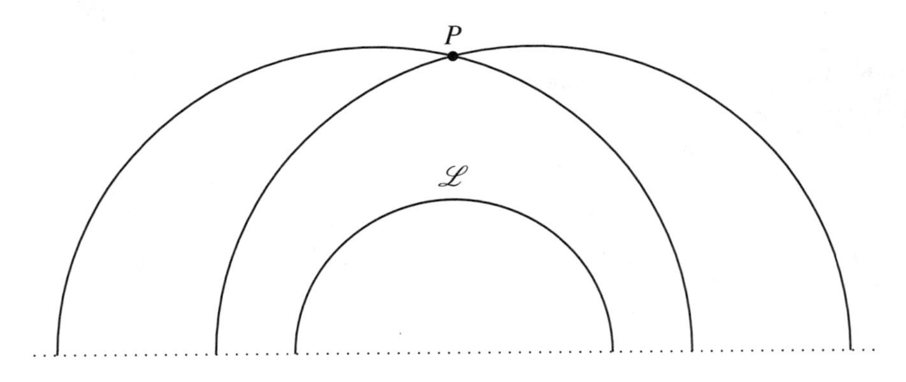

Figure 8.1: Failure of the parallel axiom for non-Euclidean "lines"

The dotted line represents the real line $y = 0$ so the half plane $y > 0$ consists of the points strictly above it. The semicircle $\mathscr{L}$ is one "line" in the half plane, and the two semicircles passing through the point P clearly do not meet $\mathscr{L}$, so they are two "parallels" of $\mathscr{L}$. In the remainder of this chapter we explain in more detail why these semicircles should be regarded as "lines," and why they satisfy all of Euclid's axioms for lines except the parallel axiom.

Thus, the complex half plane not only allows us to visualize the geometry of the projective line; it also answers a fundamental question in the foundations of geometry by showing that *the parallel axiom does not follow from Euclid's other axioms.*

The non-Euclidean "line" through two points

One property of non-Euclidean "lines" can be established immediately. *There is a unique non-Euclidean "line" through any two points* (and hence non-Euclidean "lines" satisfy the first of Euclid's axioms).

- If the two points lie on the same vertical line $x = l$, then $x = l$ is a non-Euclidean "line" containing them. And it is the only one, because a semicircle with its endpoints on the x-axis has at most one point on each line $x = l$.

- If the two points P and Q do not lie on the same vertical line, there is a unique point R on the x-axis equidistant for both of them, namely, where the equidistant line of P and Q meets the x-axis. Then the semicircle with center R through P and Q is the unique non-Euclidean "line" through P and Q.

Exercises

One can begin to understand the geometric significance of linear fractional transformations of the half plane by studying the simplest ones, $z \mapsto z + l$ and $z \mapsto kz$ for real k and l.

8.1.1 Show that the transformations $z \mapsto z + l$ and $z \mapsto kz$ (for $k > 0$) map the upper half plane onto itself and that they map "lines" to "lines."

8.1.2 Explain why this is *not* the case when k and l are not real.

8.1.3 Show how to map the semicircle $x^2 + y^2 = 1$, $y > 0$, onto the semicircle $(x-1)^2 + y^2 = 4$, $y > 0$, by a combination of transformations $z \mapsto z + l$ and $z \mapsto kz$.

8.1.4 More generally, explain why any semicircle with center on the x-axis can be mapped onto any other by a combination of transformations $z \mapsto z + l$ and $z \mapsto kz$.

What is not yet clear is why semicircles should be regarded as "lines." Their "linelike" behavior stems from the transformation $z \mapsto 1/\bar{z}$, which we study in Section 8.2.

8.2 Complex conjugation

We know from Section 5.6 that all linear fractional transformations of $\mathbb{RP}^1$ are products of the transformations $x \mapsto x+l$, $x \mapsto kx$, and $x \mapsto 1/x$ for real constants $k \neq 0$ and l. We called these the *generating transformations* of $\mathbb{RP}^1$. The transformation $x \mapsto x+l$ obviously extends to the transformation $z \mapsto z+l$, which maps the upper half plane onto itself for any real l, but the appropriate extension of $x \mapsto kx$ is $z \mapsto kz$ only for $k > 0$, because $z \mapsto kz$ does not map the upper half plane onto itself when $k < 0$. In particular, *what is the appropriate extension of $x \mapsto -x$ to a map of the upper half plane?*

Geometrically, the answer is obvious. The transformation $x \mapsto -x$ is reflection of the line in O, so its most appropriate extension is *reflection of the half plane in the y-axis*, that is, the transformation

$$x+iy \mapsto -x+iy.$$

This transformation can be expressed more simply with the help of the *complex conjugate* $\bar{z}$, which is defined as follows. If $z = x+iy$, then $\bar{z} = x-iy$. Then the reflection of z in the y-axis is $-\bar{z}$ because

$$-\bar{z} = -\overline{(x+iy)} = -(x-iy) = -x+iy.$$

Thus, the appropriate extension of $x \mapsto -x$ is $z \mapsto -\bar{z}$ (Figure 8.2).

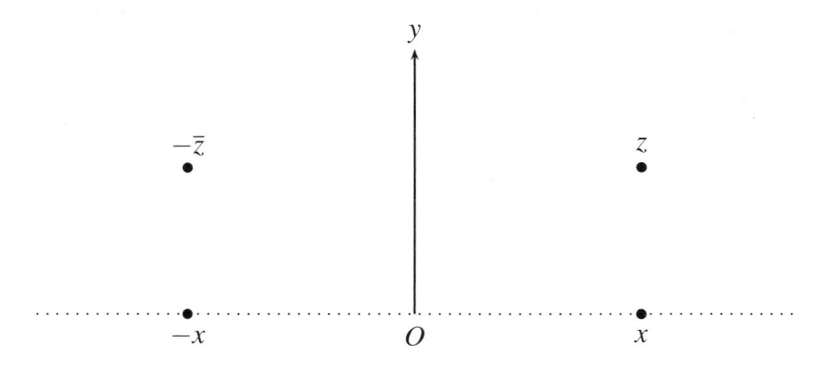

Figure 8.2: Extending reflection from the line to the half plane

More generally, the appropriate extension of $x \mapsto kx$ when $k < 0$ is $z \mapsto k\bar{z}$, the product of the reflection $z \mapsto -\bar{z}$ with the map $z \mapsto |k|z$ (dilation by factor $|k|$).

A similar problem arises when we want to extend the transformation $x \mapsto 1/x$ of $\mathbb{RP}^1$ to the half plane. The appropriate extension is not $z \mapsto 1/z$ because this transformation does not map the upper half plane onto itself. In fact, if we write z in its polar form $z = r(\cos\theta + i\sin\theta)$, then

$$\frac{1}{z} = \frac{1}{r}(\cos\theta - i\sin\theta) = \frac{1}{r}(\cos(-\theta) + i\sin(-\theta))$$

because $\cos(-\theta) = \cos\theta$ and $\sin(-\theta) = -\sin\theta$. Thus, z (at angle θ) and $1/z$ (at angle $-\theta$) have opposite slopes from O. Hence, they lie in different half planes. The appropriate extension of $x \mapsto 1/x$ is $z \mapsto 1/\bar{z}$, which sends

$$z = r(\cos\theta + i\sin\theta) \quad \text{to} \quad 1/\bar{z} = \frac{1}{r}(\cos\theta + i\sin\theta)$$

lying in the same direction θ from O (Figure 8.3). This transformation is called *reflection* (or *inversion*) *in the unit circle.*

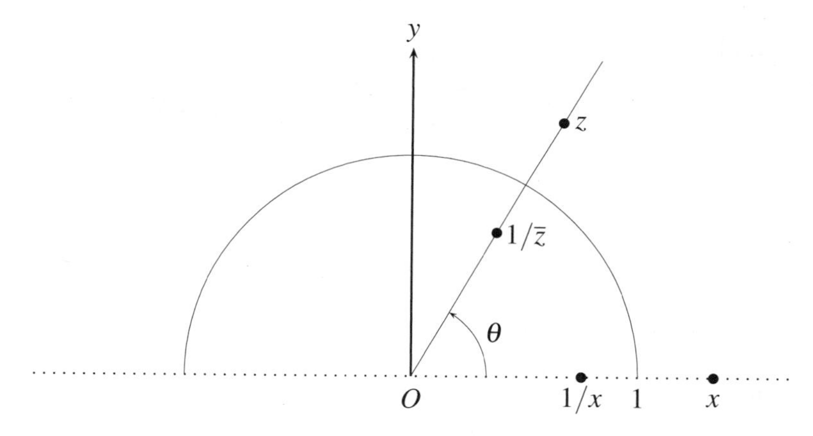

Figure 8.3: Extending inversion from the line to the half plane

Because all transformations $x \mapsto \frac{ax+b}{cx+d}$ of $\mathbb{RP}^1$ are products of $x \mapsto x + l$, $x \mapsto kx$, and $x \mapsto 1/x$, *their extensions to the half plane are products of*

- the horizontal translations $z \mapsto z + l$,
- the dilations $z \mapsto kz$ for $k > 0$,
- reflection in the y-axis $z \mapsto -\bar{z}$,
- reflection in the unit circle $z \mapsto 1/\bar{z}$.

We call these the *generating transformations of the half plane.*

Equations of non-Euclidean "lines"

Complex conjugation not only enables us to express reflection in lines and circles; it also enables us to write the equations of non-Euclidean "lines" very simply as equations in z.

- First consider "lines" that are actual Euclidean lines, namely those of the form $x = a$, where a is a real number. An arbitrary point on this line is of the form $z = a + iy$, so $\overline{z} = a - iy$, and z therefore satisfies the equation
$$z + \overline{z} = 2a. \qquad (*)$$

- Next consider "lines" that are semicircles with centers on the x-axis. If the center is c and the radius is r, then any z on the circle satisfies
$$|z - c| = r, \quad \text{or equivalently,} \quad |z - c|^2 = r^2.$$
But now notice that for any complex number $x + iy$ we have
$$|x + iy|^2 = x^2 + y^2 = (x + iy)(x - iy) = (x + iy)\overline{(x + iy)}.$$
Hence,
$$|z - c|^2 = (z - c)\overline{(z - c)} = (z - c)(\overline{z} - \overline{c})$$
and the equation $|z - c|^2 = r^2$ becomes
$$(z - c)(\overline{z} - \overline{c}) = r^2,$$
that is,
$$z\overline{z} - c\overline{z} - \overline{c}z + c\overline{c} = r^2.$$
Finally, because c is a real number, we have $\overline{c} = c$, so the equation can be written as
$$z\overline{z} - c(z + \overline{z}) + c^2 - r^2 = 0. \qquad (**)$$

The equations (*) and (**) are both of the form

$$Az\overline{z} + B(z + \overline{z}) + C = 0 \quad \text{for some} \quad A, B, C \in \mathbb{R}. \qquad (***)$$

Conversely, if A and B are not both zero, then (***) reduces to one of the equations (*) or (**) above, if it is satisfied by any points z at all.

- If $A = 0$, then (***) becomes $z+\bar{z}+C/B = 0$, which is (*) with $2a = -C/B$.
- If $A \neq 0$, then (***) becomes $z\bar{z}+(z+\bar{z})B/A+C/A = 0$, which is (**) with $c = -B/A$ and $c^2 - r^2 = C/A$ if $r^2 = B^2/C^2 - C/A \geq 0$. If $r^2 < 0$, then no points z satisfy the equation (***), because the equation is equivalent to $|z-c|^2 = r^2$ and $|z-c|^2$ is necessarily > 0.

Thus, *equations of non-Euclidean "lines" are the satisfiable equations*

$$Az\bar{z}+B(z+\bar{z})+C = 0, \quad \textit{where} \quad A,B,C \in \mathbb{R} \quad \textit{are not all zero.}$$

Exercises

I expect that most readers of this book are familiar with the complex numbers, but it still seems worthwhile to review the properties of the complex conjugate. Its role in geometric transformations may not be familiar, so we develop the basic facts from first principles.

8.2.1 Writing z_1 as $x_1 + iy_1$ and z_2 as $x_2 + iy_2$, show that $\overline{z_1 + z_2} = \overline{z_1} + \overline{z_2}$ and $\overline{z_1 z_2} = \overline{z_1}\,\overline{z_2}$.

8.2.2 Similarly, show that $\overline{1/z} = 1/\bar{z}$.

8.2.3 Deduce from Exercises 8.2.1 and 8.2.2 that, for any $a,b,c,d \in \mathbb{R}$ and $z \in \mathbb{C}$, the complex conjugate of $\frac{az+b}{cz+d}$ is $\frac{a\bar{z}+b}{c\bar{z}+d}$.

With these facts established, we are in a position to determine the extension to the half plane of each linear fractional transformation $x \mapsto \frac{ax+b}{cx+d}$ of $\mathbb{RP}^1$. What we know so far is that the extension of $x \mapsto x+l$ is $z \mapsto z+l$, the extension of $x \mapsto kx$ is $z \mapsto kz$ when $k > 0$ and $z \mapsto k\bar{z}$ when $k < 0$, and that the extension of $x \mapsto 1/x$ is $z \mapsto 1/\bar{z}$. We also know that any transformation $x \mapsto \frac{ax+b}{cx+d}$ is a product of these generating transformations. Hence, the extension of $x \mapsto \frac{ax+b}{cx+d}$ to the half plane is the product of the corresponding extensions. It seems likely that the latter product is either $z \mapsto \frac{az+b}{cz+d}$ or $z \mapsto \frac{a\bar{z}+b}{c\bar{z}+d}$, so the main problem is to decide *when* the product is $z \mapsto \frac{az+b}{cz+d}$ and when it is $z \mapsto \frac{a\bar{z}+b}{c\bar{z}+d}$.

8.2.4 Write each generating transformation of $\mathbb{RP}^1$ in the form $x \mapsto \frac{ax+b}{cx+d}$, and hence, show that those whose extension involves $\bar{z}$ are precisely those for which $ad - bc < 0$.

8.2.5 Deduce from Exercise 8.2.4 and Exercise 5.6.3 that the extension of a product, of transformations $x \mapsto \frac{a_1x+b_1}{c_1x+d_1}$ and $x \mapsto \frac{a_2x+b_2}{c_2x+d_2}$, is the product of their extensions.

8.2.6 Deduce from Exercise 8.2.5, or otherwise, that the extension of $x \mapsto \frac{ax+b}{cx+d}$ is $z \mapsto \frac{az+b}{cz+d}$ when $ad - bc > 0$ and $z \mapsto \frac{a\bar{z}+b}{c\bar{z}+d}$ otherwise.

It may seem unfortunate that the extension of $x \mapsto \frac{ax+b}{cx+d}$ is one of two different types: a function of z or a function of $\overline{z}$. However, these two algebraic types are inevitable because they reflect a geometric distinction: *The functions of z are orientation-preserving, and the functions of $\overline{z}$ are not.* In particular, *the linear fractional transformations $z \mapsto \frac{az+b}{cz+d}$ with $ad - bc > 0$ are precisely the orientation-preserving transformations of the half plane.*

8.3 Reflections and Möbius transformations

The extensions of the transformations $x \mapsto \frac{ax+b}{cx+d}$ from $\mathbb{RP}^1$ to the half plane could be called "linear fractional," but this would be confusing, because one half of them are linear fractional functions of z and the other half are linear fractional functions of $\overline{z}$. Instead they are called *Möbius transformations*, after the German mathematician August Ferdinand Möbius. In 1855, Möbius introduced a theory of transformations generated by reflections in circles, using the obvious generalization from reflection in the unit circle to reflection in an arbitrary circle. We will see below that all Möbius transformations of the half plane are products of reflections.

One advantage of the reflection idea is that it makes sense in three (or more) dimensions, where *reflection in a sphere* is meaningful but "linear fractional transformation" generally is not. It is also revealing to view the transformations of $\mathbb{RP}^1$ as the restrictions of Möbius transformations of the half plane, as this brings to light a concept of "projective reflection".

Reflection in an arbitrary circle is defined by generalizing the relationship between z and $1/\overline{z}$ shown in Figure 8.3. We say that points Q and Q' are *reflections of each other in the circle with center P and radius r* if P, Q, Q' lie in a straight line and $|PQ||PQ'| = r^2$ (Figure 8.4).

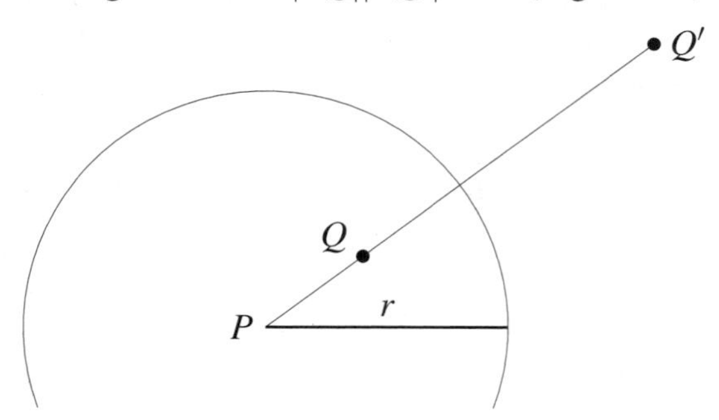

Figure 8.4: Reflection in an arbitrary circle

If the circle (or, rather, its upper half) is a non-Euclidean line, then the center P lies on the x-axis, and reflection in this circle can be composed from generating transformations of the half plane as follows:

- translate P to O,
- reduce the radius to 1 by dilating by $1/r$,
- reflect in the unit circle,
- restore the radius to r by dilating by r,
- translate the center from O back to P.

Likewise, reflection in an arbitrary vertical line, say $x = a$, can be composed from generating transformations of the half plane as follows:

- translate the line $x = a$ to the y-axis,
- reflect in the y-axis,
- translate the y-axis to the line $x = a$.

Thus, *all reflections in non-Euclidean lines are products of generating transformations of the half plane.*

Conversely, we now show that *every generating transformation of the half plane is a product of reflections* (and hence so is every transformation of the half plane). The generating transformations $z \mapsto -\bar{z}$ and $z \mapsto 1/\bar{z}$ are reflections by definition, so it remains to deal with the remaining generating transformations.

- the horizontal translation $z \mapsto z + l$: this is a Euclidean translation, and it is the product of reflections in the lines $x = 0$ and $x = l/2$.
- the dilation $z \mapsto kz$, where $k > 0$: this is the product of the reflection $z \mapsto 1/\bar{z}$ in the unit circle and the map $z \mapsto k/\bar{z}$, which is reflection in the circle with center O and radius $\sqrt{k}$. □

It should be mentioned that ordinary reflection—reflection in a straight line—is the limiting case of reflection in a circle obtained by letting P and r tend to infinity in such a way that the circle tends to a straight line. Because Euclidean lines are the fixed point sets of ordinary reflections, it is natural that the "lines" of the half plane should be the fixed point sets of its "reflections."

Projective reflections

Looking back from the half plane to its boundary line $\mathbb{RP}^1$, we realize that we now know more about projective transformations of the line than we did before. *Any projective transformation of* $\mathbb{RP}^1$ *is a product of projective reflections*, where a *projective reflection* is the restriction, to $\mathbb{RP}^1$, of a reflection of the half plane.

There is a "three reflections theorem" for $\mathbb{RP}^1$, analogous to the three reflections theorem for isometries of the Euclidean plane (Section 3.7). This follows from a three reflections theorem for the half plane, similar to the one for the Euclidean plane, that we will prove in Section 8.8.

Exercises

The simplest reflections of $\mathbb{RP}^1$ are ordinary reflection in O, $x \mapsto -x$, and the restriction of reflection in the unit circle, $x \mapsto 1/x$. The map $x \mapsto 1/x$ might be called "reflection in the point-pair $\{-1,1\}$," and it generalizes to "reflection in the point-pair $\{a,b\}$." (A point-pair $\{a,b\}$ is a "0-dimensional sphere," because it consists of the points at constant distance $(b-a)/2$ from the "center" $(a+b)/2$.)

8.3.1 Write down the formula for ordinary reflection in the point $x = a$.

8.3.2 Explain why the map $x \mapsto c^2/x$ is reflection in the point-pair $\{-c,c\}$.

8.3.3 Using Exercise 8.3.2, or otherwise, show that reflection in the point-pair $\{a,b\}$ is given by the linear fractional function

$$f(x) = \frac{x(a+b)-2ab}{2x-(a+b)}.$$

8.3.4 Show that, as $b \to \infty$, the function for reflection in the point-pair $\{a,b\}$ tends to the function for ordinary reflection in the point $x = a$.

8.4 Preserving non-Euclidean lines

We have now extended the projective transformations $x \mapsto \frac{ax+b}{cx+d}$ of $\mathbb{RP}^1$ to Möbius transformations $z \mapsto \frac{az+b}{cz+d}$ or $z \mapsto \frac{a\bar{z}+b}{c\bar{z}+d}$ of the half plane, but are Möbius transformations of the half plane any easier to understand? We intend to show that they are, by showing that they have more easily visible invariants than the transformations of $\mathbb{RP}^1$. First we show the invariance of *non-Euclidean lines*, which we now define officially as the vertical lines $x =$ constant and the semicircles with centers on the x-axis.

Each Möbius transformation of the half plane maps non-Euclidean lines to non-Euclidean lines.

For the generating transformations $z \mapsto z+l$, $z \mapsto kz$ for $k > 0$, and $z \mapsto -\bar{z}$, this is easy to see. Each of these transformations sends vertical lines to vertical lines, circles to circles, and the x-axis to the x-axis because:

- $z \mapsto z+l$ is a horizontal translation of the half plane.
- $z \mapsto kz$ with $k > 0$ is a dilation of the half plane by k.
- $z \mapsto -\bar{z}$ is the Euclidean reflection of the half plane in the y-axis.

Thus, any product of the three transformations just listed sends vertical lines to vertical lines and semicircles with centers on the x-axis to semicircles with centers on the x-axis. Hence, *all products of the transformations* $z \mapsto z+l$, $z \mapsto kz$ *for* $k > 0$, *and* $z \mapsto -\bar{z}$ *preserve non-Euclidean lines.*

To show that all Möbius transformations preserve non-Euclidean lines, it therefore remains to show that *reflection in the unit circle,* $z \mapsto 1/\bar{z}$, *preserves non-Euclidean lines.* This is less obvious, because reflection in a circle can send a vertical line to a semicircle and vice versa. We prove that non-Euclidean lines are preserved by using their equations (***) derived in Section 8.3.

Given a non-Euclidean line, whose points z satisfy an equation

$$Az\bar{z} + B(z+\bar{z}) + C = 0 \quad \text{for some} \quad A, B, C \in \mathbb{R}, \tag{***}$$

we wish to find the equation satisfied by the points of its reflection in the unit circle. These are the points $w = 1/\bar{z}$, so we seek the equation satisfied by w. The required equation is likely to involve $\overline{w} = 1/z$ as well, so we are looking for an equation connecting $1/z$ and $1/\bar{z}$. Such an equation is easy to find: just divide the equation (***) by $z\bar{z}$. Division yields the equation

$$A + B\left(\frac{1}{\bar{z}} + \frac{1}{z}\right) + \frac{C}{z\bar{z}} = 0,$$

that is,

$$Cw\overline{w} + B(w + \overline{w}) + A = 0. \tag{****}$$

Equation (****) is satisfied by the reflections $w = 1/\bar{z}$ of the points z satisfying (***), and (****) has the same form as (***), because $A, B, C \in \mathbb{R}$. Hence, (****) also represents a non-Euclidean line. □

Exercises

An example in which reflection in the unit circle sends a vertical line to a semicircle is shown in Figure 8.5.

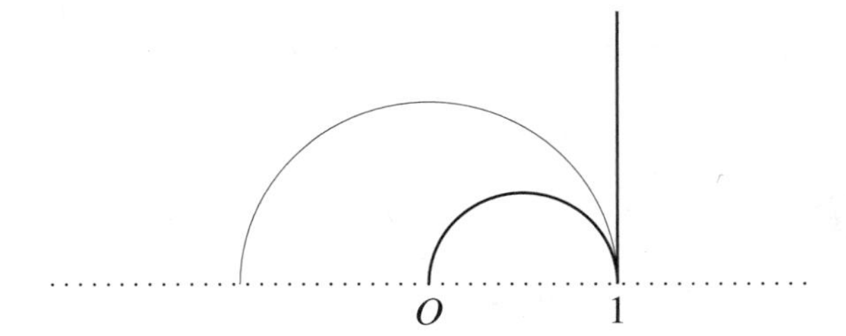

Figure 8.5: Reflection of the line $x = 1$

8.4.1 Give intuitive reasons why the reflection of the line $x = 1$ in the unit circle should have one end at 1 on the x-axis and the other end at O.

8.4.2 Show that the line $x = 1$ has equation $z + \overline{z} = 2$, and that its reflection in the unit circle has equation $w + \overline{w} = 2w\overline{w}$.

8.4.3 Verify that $w + \overline{w} = 2w\overline{w}$ is the equation of the semicircle with ends O and 1 on the x-axis.

8.5 Preserving angle

Next to non-Euclidean lines, the most visible invariant of Möbius transformations is *angle*. Because non-Euclidean lines are not necessarily straight, the angle between two of them is really the angle between their tangents at the point of intersection. Nevertheless, it is easy to see the angle between non-Euclidean lines. Figure 8.6 shows an example.

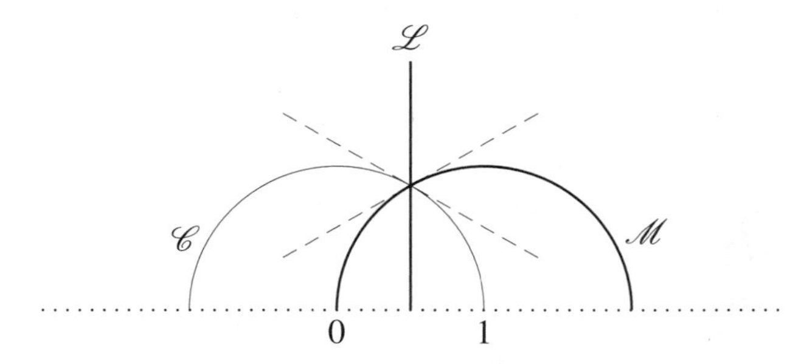

Figure 8.6: Some non-Euclidean lines and the angles between them

The three non-Euclidean lines are the unit circle $\mathscr{C}$, the vertical $\mathscr{L}$ where $x = 1/2$, and its reflection $\mathscr{M}$ in the unit circle, which happens to be the semicircle with endpoints 0 and 2 on the x-axis.

At the point where the three non-Euclidean lines meet, they divide the space around the point into six equal angles, so each angle is $2\pi/6 = \pi/3$. This equality is confirmed by the tangents, which are shown as dashed lines. Notice that any two of $\mathscr{C}$, $\mathscr{L}$, and $\mathscr{M}$ are reflections of each other in the third non-Euclidean line, so the figure shows numerous instances of an angle equal to its reflection. To show that *any* angle is preserved by *any* Möbius transformation, we look once again at the properties of the generating transformations.

The effect of Möbius transformations

The Möbius transformations $z \mapsto z + l$ and $z \mapsto -\bar{z}$ are Euclidean isometries; hence, they certainly preserve angle (along with length, area, and so on). The Möbius transformations $z \mapsto kz$ for $k > 0$ are dilations; hence, they too preserve angle. Thus, *it suffices to prove that angle is preserved by the remaining generator of Möbius transformations: reflection in the unit circle*, $z \mapsto 1/\bar{z}$. The latter transformation is the composite of $z \mapsto -\bar{z}$ and $z \mapsto -1/z$, so it suffices in turn to prove that $z \mapsto -1/z$ preserves angle.

We therefore concentrate our attention on the Möbius transformation $z \mapsto -1/z$. This transformation does not in general preserve Euclidean lines, because it may map them to circles. Thus, we need to be aware that "angle" generally means the angle between curves and hence the angle between the tangents. However, we can avoid computing the position of tangents by taking the *infinitesimal* view of angle. That is, we study what becomes of the direction between two points, z and $z + \Delta z$, when we send them to $-1/z$ and $-1/(z + \Delta z)$, respectively, and let Δz tend to zero.

If Δz is the point at distance ε from O in direction θ, then

$$\Delta z = \varepsilon(\cos\theta + i\sin\theta),$$

because $\cos\theta + i\sin\theta$ is the point at distance 1 from O in direction θ. It follows that the point at distance ε from z in direction θ is

$$z + \Delta z = z + \varepsilon(\cos\theta + i\sin\theta),$$

and that the point $z + \Delta z$ tends to z in the constant direction θ as ε tends to zero.

The difference between the image points $-1/(z+\Delta z)$ and $-1/z$ is therefore

$$\frac{1}{z}-\frac{1}{z+\varepsilon(\cos\theta+i\sin\theta)}=\frac{z+\varepsilon(\cos\theta+i\sin\theta)-z}{z(z+\varepsilon(\cos\theta+i\sin\theta))}$$
$$=\frac{\varepsilon(\cos\theta+i\sin\theta)}{z(z+\varepsilon(\cos\theta+i\sin\theta))}.$$

Now as ε tends to zero, this difference is ever more closely approximated by

$$\frac{\varepsilon(\cos\theta+i\sin\theta)}{z^2}.$$

To be precise, *the direction from* $-1/z$ *to* $-1/(z+\Delta z)$ *tends to the direction of* $\varepsilon(\cos\theta+i\sin\theta)z^{-2}$, *which is* θ+constant. The constant is the argument (angle) of z^{-2}, recalling from Section 4.7 that the argument of a product of complex numbers is the sum of their arguments.

The angle between two smooth curves meeting at z (approximated by the difference in directions from z to points $z+\Delta_1 z$ and $z+\Delta_2 z$ on the respective curves) is therefore the angle between the images of these curves under the map $z\mapsto -1/z$. This is because a smooth curve is one for which the direction from z to $z+\Delta z$ tends to a constant θ as $z+\Delta z$ tends to z along the curve. Non-Euclidean lines are smooth, so the angle between them is preserved by the transformation $z\mapsto -1/z$, as required. □

Tilings of the half plane

If one takes a triangle with angles π/p, π/q, π/r, for some natural numbers p,q,r, then any reflection of that triangle will have angles π/p, π/q, π/r. Reflecting the reflections causes the space around each vertex to be exactly filled with corners of triangles. For example, the space around the vertex of angle π/p becomes filled with $2p$ corners of angle π/p. In fact, the *whole half plane* becomes filled, or *tiled*, by copies of the original triangle. An example is shown in Figure 8.7, where the basic tile has angles $\pi/2$, $\pi/3$, and $\pi/7$.

Notice that the angle sum $\pi/2+\pi/3+\pi/7$ is less than π. In fact, the angle sum of any triangle bounded by non-Euclidean lines is less than π, and *the quantity* $(\pi-\text{angle sum})$ *is proportional to the area of the triangle.* This elegant result is less surprising when one learns that the area of spherical triangle is also proportional to $\pi-$ angle sum (see exercises below).

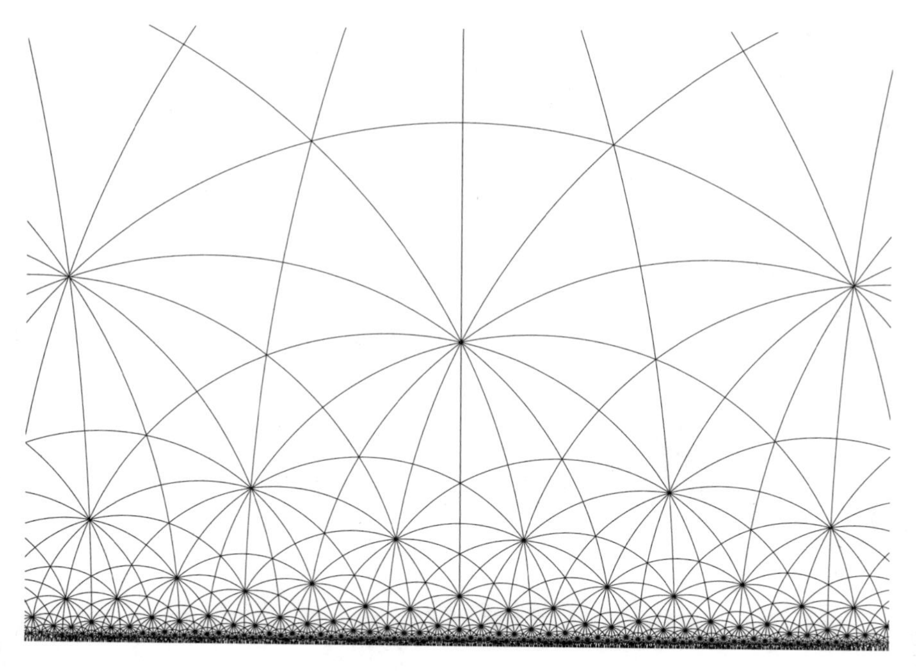

Figure 8.7: Tiling by repeated reflections

However, it does reveal a limitation in the half plane view of non-Euclidean geometry: All the triangles in Figure 8.7 have equal non-Euclidean area, but they certainly do not look equal!

One should think of the half plane as a kind of "perspective view" of the non-Euclidean plane with the x-axis as a horizon. The x-axis is infinitely distant, because there are infinitely many identical triangles between any point of the half plane and the x-axis. In this respect, the half plane is like a perspective view of a Euclidean tiled floor, except that ordinary perspective preserves straightness and distorts angle, whereas this "non-Euclidean perspective" distorts straightness and preserves angle. There are other views of the non-Euclidean plane that make non-Euclidean lines look straight (see Section 8.9), but any such view has a curved horizon!

Another way in which a tiling of the half plane resembles a perspective view is that one can estimate the length of a line by counting the numbers of tiles that lie along it. There is indeed a non-Euclidean measure of distance that is invariant under Möbius transformations, and we will see exactly what it is in Section 8.6.

Exercises

The English mathematician Thomas Harriot discovered that the area of a spherical triangle is proportional to (angle sum $-\ \pi$) in 1603. His argument is based on the two views of a spherical triangle shown in Figure 8.8.

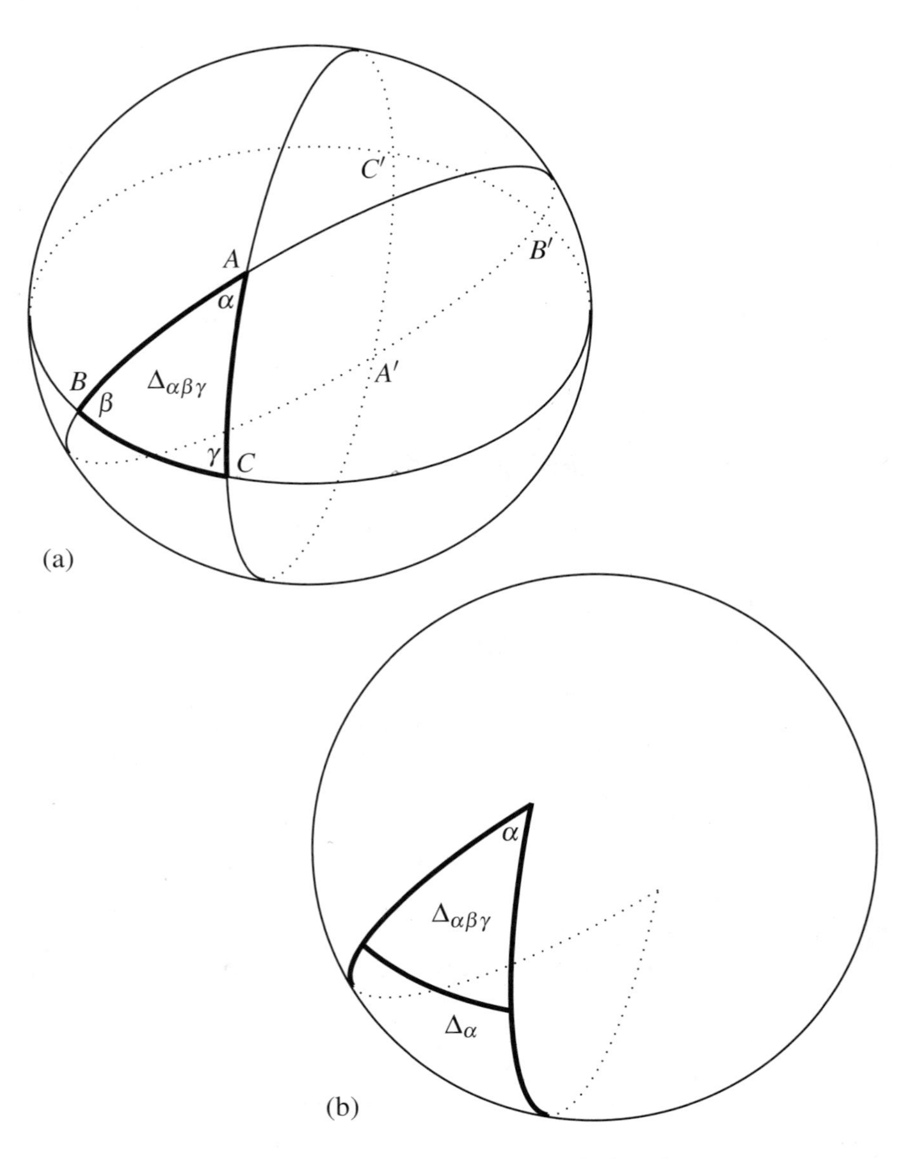

Figure 8.8: Area of a spherical triangle

View (a) shows all sides of the spherical triangle extended to great circles. These divide the sphere into eight spherical triangles, which are obviously congruent in antipodal pairs. View (b) shows the result of extending two sides, which is a "slice" of the sphere with area proportional to the angle at its two ends.

8.5.1 Letting the area of the triangle with angles α,β,γ be $\Delta_{\alpha\beta\gamma}$, and letting the areas of the other triangles be $\Delta_\alpha,\Delta_\beta,\Delta_\gamma$ as shown in view (a), prove that

$$2\left(\Delta_{\alpha\beta\gamma}+\Delta_\alpha+\Delta_\beta+\Delta_\gamma\right)=\text{area of sphere, call it } A. \qquad (1)$$

8.5.2 Use view (b) to explain why

$$\Delta_{\alpha\beta\gamma}+\Delta_\alpha=\frac{\alpha}{2\pi}A,\quad \Delta_{\alpha\beta\gamma}+\Delta_\beta=\frac{\beta}{2\pi}A,\quad \Delta_{\alpha\beta\gamma}+\Delta_\gamma=\frac{\gamma}{2\pi}A.$$

8.5.3 Deduce from Exercise 8.5.2 that

$$3\Delta_{\alpha\beta\gamma}+\Delta_\alpha+\Delta_\beta+\Delta_\gamma=\frac{\alpha+\beta+\gamma}{2\pi}A \qquad (2)$$

8.5.4 Deduce from equations (1) and (2) that $4\Delta_{\alpha\beta\gamma}=\frac{\alpha+\beta+\gamma-\pi}{\pi}A$, and hence that

$$\Delta_{\alpha\beta\gamma}=\text{constant}\times(\alpha+\beta+\gamma-\pi).$$

8.5.5 Using a formula for the area of the sphere, show that $\Delta_{\alpha\beta\gamma}=\alpha+\beta+\gamma-\pi$ on a sphere of radius 1.

8.6 Non-Euclidean distance

So far we have found invariants of Möbius transformations by geometrically inspired guesses that can be confirmed by calculations with linear fractional functions. But still up our sleeve is the cross-ratio card, which carries the fundamental invariant of linear fractional transformations, and to find out what non-Euclidean distance is we finally have to play it.

We know from Section 5.7 that the cross-ratio is invariant under the transformations $x\mapsto x+l$ and $x\mapsto kx$, and exactly the same calculations apply to $z\mapsto z+l$ and $z\mapsto kz$. It is invariant under $z\mapsto -1/z$, as can be shown by a calculation similar to, but shorter than, that given in Section 5.7 for $x\mapsto 1/x$. However, it is *not* generally invariant under the Möbius transformation $z\mapsto -\bar{z}$, because this replaces the cross-ratio by its complex conjugate. We can only say that *Möbius transformations either leave the cross-ratio invariant or change it to its complex conjugate.*

Luckily, this does not matter, because we are interested in the cross-ratio only when the four points lie on a non-Euclidean line. It turns out that *the cross-ratio of four points on a non-Euclidean line is real, and hence equal to its own complex conjugate.*

This is obvious when the points are pi, qi, ri, si on the upper y-axis, because, in this case, the cross-ratio equals the real number $\frac{(r-p)(s-q)}{(r-q)(s-p)}$ by cancellation of the i factors. It follows for any other non-Euclidean line $\mathscr{L}$ by mapping the upper y-axis onto $\mathscr{L}$ by a Möbius transformation.

- If $\mathscr{L}$ is another vertical line $x = l$, we map the upper y-axis to $\mathscr{L}$ by $z \mapsto z + l$.
- If $\mathscr{L}$ is a semicircle with center on the x-axis, we first map the upper y-axis to $x = 1$ by $z \mapsto z + 1$, and then to the semicircle with ends at 0 and 1 by $z \mapsto 1/\bar{z}$. Finally we map this semicircle to $\mathscr{L}$ by dilating it to the radius of $\mathscr{L}$ and then translating its center to the center of $\mathscr{L}$. □

We know from the previous section that the transformations $z \mapsto z + l$, $z \mapsto kz$ for $k > 0$, $z \mapsto -\bar{z}$, and $z \mapsto -1/z$ generate all Möbius transformations, so we have now proved that *the cross-ratio of any four points on a non-Euclidean line is preserved by Möbius transformations.*

So far, so good, but distance is a function of two points, not four. If the cross-ratio is going to help us define distance, we need to specialize it to a function of two variables.

One of the beauties of a non-Euclidean line is that it lies between two endpoints. The non-Euclidean line represented by the upper y-axis, for example, consists of the points between 0 and ∞. The endpoints are not points of the line, but it is meaningful to include them in a cross-ratio, because Möbius transformations apply to all complex numbers, and ∞. If we take 0 and ∞ as the third and fourth members of the quadruple pi, qi, ri, si on the upper y-axis, then the cross-ratio of this quadruple simplifies as follows:

$$\begin{aligned}
\frac{(r-p)(s-q)}{(r-q)(s-p)} &= \frac{(r-p)(1-q/s)}{(r-q)(1-p/s)} && \text{dividing top and bottom by } s \\
&= \frac{r-p}{r-q} && \text{because } s = \infty \text{ and } 1/\infty = 0 \\
&= \frac{p}{q} && \text{because } r = 0.
\end{aligned}$$

Any Möbius transformation of the upper y-axis sends endpoints to endpoints, as one can see from the generating transformations, but it is possible for 0 and ∞ to be exchanged. If $r = \infty$ and $s = 0$, we find that the cross-ratio of pi, qi, ri, si is q/p, not p/q. Thus, q/p is not an invariant of Möbius transformations of the upper y-axis, *but* $|\log \frac{q}{p}|$ *is*, because $\log \frac{p}{q} = -\log \frac{q}{p}$ for any $p, q > 0$.

This prompts us to make the following definition.

Distance on the upper y-axis. The *non-Euclidean distance* $\text{ndist}(pi, qi)$ between points pi and qi on the upper y-axis is $|\log \frac{q}{p}|$.

This definition of distance is appropriate for two reasons:

- As already shown, non-Euclidean distance on the upper y-axis is invariant under all Möbius transformations.
- Non-Euclidean distance is *additive*. That is, if pi, qi, ri lie on the upper y-axis in that order, then

$$\text{ndist}(pi, ri) = \text{ndist}(pi, qi) + \text{ndist}(qi, ri).$$

This is because

$$\left|\log \frac{r}{p}\right| = \left|\log \frac{q}{p}\frac{r}{q}\right| = \left|\log \frac{q}{p} + \log \frac{r}{q}\right| = \left|\log \frac{q}{p}\right| + \left|\log \frac{r}{q}\right|$$

by the additive property of the logarithm function.

It follows from this definition that the infinity of points $2^n i$, for integers n, are *equally spaced* along the upper y-axis, in the sense of non-Euclidean distance. The faces shown in Figure 8.9 are of equal size in this sense. The upper y-axis is not only infinite in the upward direction, but also in the downward direction. There is infinite non-Euclidean distance between any of its points and the x-axis. Thus, the upper y-axis satisfies Euclid's second axiom for "lines": Any segment of it can be "extended indefinitely."

Having defined non-Euclidean distance on the upper y-axis, we can use the axis as a "ruler" to measure the distance between two points in the upper half plane. Given any two points u and v, we find the unique non-Euclidean line $\mathscr{L}$ through u and v as described in Section 8.1, and then map $\mathscr{L}$ onto the upper y-axis by a Möbius transformation f as described (in reverse) in the first part of this section. We take the non-Euclidean distance from u to v to be the non-Euclidean distance from $f(u)$ to $f(v)$, namely $\text{ndist}(f(u), f(v))$.

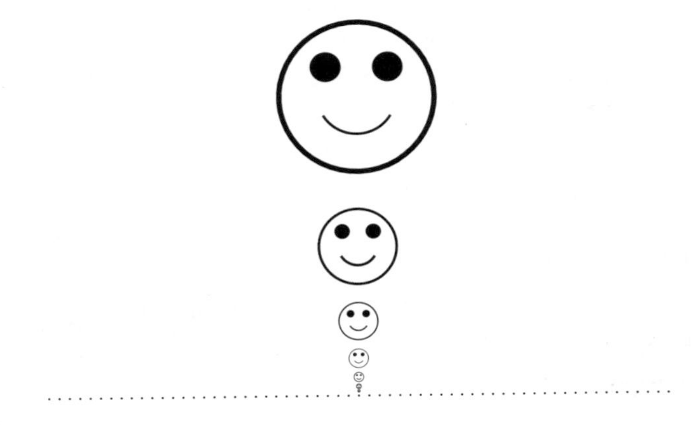

Figure 8.9: Faces of equal non-Euclidean size

The quantity $\text{ndist}(f(u), f(v))$ does not depend on the Möbius transformation f used to map $\mathscr{L}$ onto the upper y-axis. If g is another Möbius transformation mapping $\mathscr{L}$ onto the upper y-axis, then fg^{-1} is a Möbius transformation that maps the upper y-axis onto itself and sends the points $g(u)$ and $g(v)$ to $f(u)$ and $f(v)$, respectively. Hence,

$$\text{ndist}(g(u), g(v)) = \text{ndist}(f(u), f(v))$$

by the invariance of non-Euclidean distance on the upper y-axis under Möbius transformations.

The hidden geometry of the projective line

As we mentioned in Section 7.1, Klein associated a "geometry" with each group of transformations. We have set up the group of transformations of the half plane to be isomorphic to the group of transformations of $\mathbb{RP}^1$. Hence, the half plane and $\mathbb{RP}^1$ have *isomorphic geometries* in the sense of Klein, even though they seem very different. Indeed, we transferred geometry from $\mathbb{RP}^1$ to the half plane mainly because of the difference: Geometry is much more *visible* in the half plane.

Figure 8.7 is one illustration of this, and Figure 8.10 is another—a regular tiling of the half plane by fish that are congruent in the sense of non-Euclidean length. Figure 8.10 is essentially the picture *Circle Limit I*, by M. C. Escher, but mapped to the half plane by the transformation

$$z \mapsto \frac{1 - zi}{z - i}.$$

Figure 8.10: Half plane version of Escher's *Circle Limit I*

By restricting Möbius transformations to the boundary of the half plane, half plane geometry can be compressed into the geometry of $\mathbb{RP}^1$, even though $\mathbb{RP}^1$ has no concepts of length or angle. Conversely, length and angle emerge when $\mathbb{RP}^1$ is expanded to the half plane.

Exercises

We can now confirm the impression given by Figure 8.7, that each non-Euclidean line is infinite in both directions, as demanded by Euclid's second axiom.

8.6.1 Show that the y-axis, and hence any non-Euclidean line, can be divided into infinitely many segments of equal non-Euclidean length.

8.6.2 Find a Möbius transformation sending $0, \infty$ to $-1, 1$, respectively, and hence mapping the y-axis onto the unit semicircle.

8.6.3 Using the transformation found in Exercise 8.6.2, find an infinite sequence of points on the unit semicircle that are equally spaced in the sense of non-Euclidean length.

Supposing that the equal faces shown in Figure 8.9 have non-Euclidean width ε, which can be as small as we please, we can draw some interesting conclusions about the non-Euclidean distance between non-Euclidean lines.

8.6.4 Show that the non-Euclidean distance between the lines $x = 0$ and $x = 1$ tends to zero as y tends to ∞.

8.6.5 Show that the Möbius transformation $z \mapsto 2/(1-z)$ sends the unit circle and the line $x = 1$ to the lines $x = 1$ and $x = 0$, respectively.

8.6.6 Deduce from Exercise 8.6.5 that the non-Euclidean distance between the unit circle and the line $x = 1$ tends to zero as these non-Euclidean lines approach the x-axis.

8.7 Non-Euclidean translations and rotations

Like the Euclidean plane, the half plane has isometries called *translations* and *rotations*, which are products of two reflections. Their nature depends on whether the lines of reflection meet or have a common end.

A translation is the product of reflections in non-Euclidean lines that do not meet and do not have a common end. A simple example is $z \mapsto 2z$, which is the product of reflections in the circles with center 0 and radii 1 and $\sqrt{2}$. This translation maps each face in Figure 8.9 to the one above it. Any non-Euclidean translation maps a unique non-Euclidean line, called the *translation axis*, into itself. Also mapped into themselves are the curves at constant non-Euclidean distance from the translation axis, which (for distance > 0) are *not* non-Euclidean lines. For $z \mapsto 2z$, the translation axis is the y-axis and the equidistant curves are the Euclidean lines $y = ax$. Each non-Euclidean line perpendicular to the translation axis is mapped onto another such line.

Figure 8.11 shows the translation axis, two equidistant curves (in gray), and some of their perpendiculars (on the left when the axis is vertical, and on the right when it is not). Notice that the equidistant curves in general are Euclidean circles passing through the two ends of the translation axis. The translation moves each non-Euclidean perpendicular to the next.

The product of reflections in two non-Euclidean lines that meet at a point P is a *non-Euclidean rotation* about P. The point P remains fixed and points at non-Euclidean distance r from P remain at non-Euclidean distance r from P, since reflection is a non-Euclidean isometry. Hence, these points move on a *non-Euclidean circle of radius r*. It turns out that a non-Euclidean circle is a Euclidean circle, although its non-Euclidean

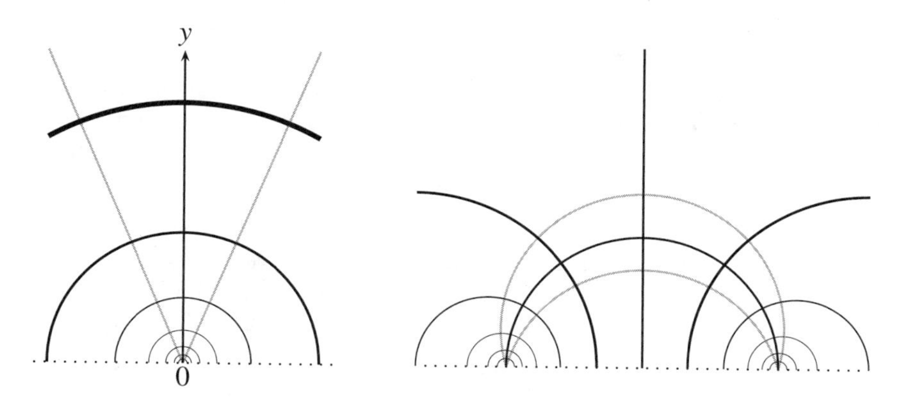

Figure 8.11: Non-Euclidean translations

center (the point at constant non-Euclidean distance from all its points) is not its Euclidean center.

For example, if we take the product of the reflection $z \mapsto -\overline{z}$ in the y-axis with the reflection $z \mapsto -1/\overline{z}$ in the unit circle, the result is a rotation through angle π about the point i where these two non-Euclidean lines meet. More generally, if we have two non-Euclidean lines through P meeting at angle θ, then the product of reflections in these lines is a rotation about P through angle 2θ. Figure 8.12 shows four non-Euclidean lines through i and two non-Euclidean circles (in gray) with non-Euclidean center at i. A rotation of $\pi/4$ about i moves each non-Euclidean line to the next and maps each circle into itself.

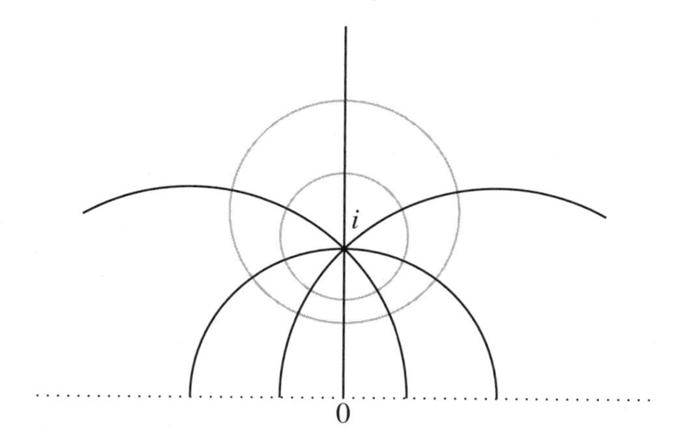

Figure 8.12: A non-Euclidean rotation about i

A limiting case of rotation is where the two lines of reflection do not meet in the half plane, but have a common end P on the boundary $\mathbb{R} \cup \{\infty\}$ at infinity. Here P is a fixed point, each non-Euclidean line ending at P is moved to another line ending at P, and each curve perpendicular to all these lines is mapped onto itself. This kind of isometry is called a *limit rotation*, and each curve mapped onto itself is called a *limit circle* or *horocycle*.

The simplest example is the Euclidean horizontal translation $z \mapsto z+1$, which is the product of reflections in the vertical lines $x=0$ and $x=1/2$. Each vertical line $x=a$ is mapped to the line $x=a+1$, and each horizontal line $y=b$ is mapped onto itself. Thus, the horizontal lines $y=b$, which we know are *not* non-Euclidean lines, are limit circles.

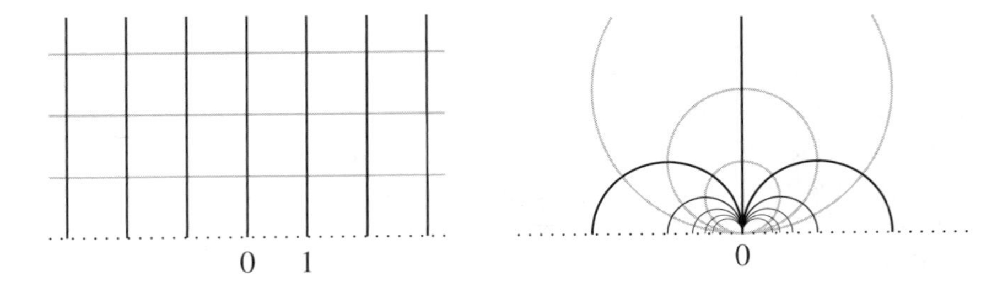

Figure 8.13: Limit rotations

Like equidistant curves, limit circles can be Euclidean lines, but generally they are Euclidean circles. Figure 8.13 shows the exceptional case $z \mapsto z+1$, where the limit circles are the Euclidean horizontal lines (in gray), and the typical case $z \mapsto z/(1-z)$, where the limit circles are the gray circles tangential to the boundary at the fixed point $z=0$.

As in the previous pictures, the isometry moves each non-Euclidean line to the next, and maps each gray curve onto itself.

Exercises

8.7.1 Check that the product of reflections in the y-axis and the unit circle is $z \mapsto -1/z$, and that i is the fixed point of this map.

8.7.2 Show also that $z \mapsto -1/z$ maps each circle of the form $|z-ti| = \sqrt{t^2-1}$ onto itself.

The limit rotation $z \mapsto z/(1-z)$ above is obtained by moving the limit rotation $z \mapsto z+1$ about ∞ to a limit rotation about 0 with the help of the rotation $z \mapsto -1/z$ that exchanges 0 and ∞.

8.7.3 If $f(z) = z+1$ and $g(z) = -1/z$, show that $gfg^{-1}(z) = z/(1-z)$.

8.7.4 Describe in words what g^{-1}, f, g in succession do to the half plane, and hence explain geometrically why gfg^{-1} has fixed point 0.

8.8 Three reflections or two involutions

It is possible to prove that each isometry of the half plane is the product of three reflections, following much the same approach as was used in Section 3.7 to prove the three reflections theorem for the Euclidean plane. The details of this approach are worked out in my book *Geometry of Surfaces*.

However, our approach to isometries of the Euclidean plane began with a definition of Euclidean distance; we then had to *find* the transformations that leave it invariant. Here we know the isometries of the half plane—the Möbius transformations—so the only problem is to express them as products in some simple way. To do this, we can interpret Möbius transformations on $\mathbb{RP}^1$, and exploit known theorems of projective geometry. Surprisingly, there is a theorem about $\mathbb{RP}^1$ that goes one better than the three reflections theorem, namely the *two involutions theorem* from Veblen and Young's 1910 book *Projective Geometry*, p. 223.

An *involution* is a transformation f such that f^2 is the identity. Thus, the involutions include the reflections, but some other transformations as well, such as the function $x \mapsto -1/x$, which (when extended to the half plane) represents a half turn about the point i. The name "involution" is one of many terms introduced into projective geometry by Desargues, and it is the only one that has stuck.

To pave the way for the two involutions theorem (and the three reflections theorem that follows from it), we first note three consequences of the results in Section 5.8 about transformations of $\mathbb{RP}^1$.

- *Any four points $p, q, r, s \in \mathbb{RP}^1$ can be mapped to q, p, s, r, respectively, by a linear fractional transformation.*

  Notice that $[p, q : r, s] = [q, p : s, r]$ because

$$\frac{(r-p)(s-q)}{(r-q)(s-p)} = \frac{(s-q)(r-p)}{(s-p)(r-q)}.$$

  Hence, by the "criterion for four-point maps" in Section 5.8, there is a linear fractional f mapping p, q, r, s to q, p, s, r, respectively.

- *If g is a linear fractional transformation that exchanges two points, then g is an involution.*

  Suppose that p and q are two points with $g(p) = q$ and $g(q) = p$. Let r be another point, not fixed by g, and suppose that $g(r) = s$. Because any linear fractional function is one-to-one, it follows that p, q, r, s are different. Hence, by the previous result, there is a linear fractional f mapping p, q, r, s to q, p, s, r, respectively.

  Because f agrees with g on the three points p, q, r, the functions f and g are identical by the "uniqueness of three-point maps" in Section 5.8. For any nonfixed point r of g, we therefore have

  $$g^2(r) = g(s) = f(s) = r,$$

  and if r is a fixed point, then $g^2(r) = r$ obviously. Hence, $g^2(x) = x$ for any $x \in \mathbb{RP}^1$, and so g is an involution.

- *For any three points p, q, r, there is an involution that exchanges p, q and fixes r.*

  By "existence of three-point maps" from Section 5.8, there is a linear fractional function g that sends p, q, r to q, p, r, respectively. Thus, g fixes r, and because it exchanges p and q, it is an involution by the previous result.

Two involutions theorem. *Any linear fractional transformation h of $\mathbb{RP}^1$ is the product of two involutions.*

If $h =$ identity, then $h =$ identity $\cdot$ identity, which is the product of two involutions. If not, let p be a point not fixed by h, so

$$h(p) = r \neq p,$$

and let $h(r) = q$. Then $q \neq r$, because h^{-1} is also a linear fractional transformation and hence one-to-one. If $q = p$, then h exchanges p and r. Hence, h is itself an involution by the second result above.

We can therefore assume that p, q, r are three different points; in which case, the third result above gives a linear fractional involution f such that

$$f(p) = q, \quad f(q) = p, \quad f(r) = r.$$

Also, fh exchanges the two points p and r because

$$fh(p) = f(r) = r, \quad fh(r) = f(q) = p.$$

Thus, fh is an involution too; call it g. Finally, applying f^{-1} to both sides of $fh = g$, we get

$$h = f^{-1}g.$$

So h is the product of two involutions, $f^{-1} = f$ and g, as required. □

We now consider Möbius transformations of the half plane, each of which is the unique extension of a linear fractional transformation of $\mathbb{RP}^1$. Such a function is determined by its values at three points on $\mathbb{RP}^1$, by "uniqueness of three-point maps." We use the same letter for a linear fractional transformation of $\mathbb{RP}^1$ and its extension to a Möbius transformation of the half plane, and we systematically use the fact that Möbius transformations preserve non-Euclidean lines and angles.

Three reflections theorem. *Any Möbius transformation of the half plane is the product of at most three reflections.*

The involution f in the proof above, which exchanges p, q and fixes r, necessarily maps the non-Euclidean line $\mathscr{L}$ from p to q into itself. Points of $\mathscr{L}$ near the end p are sent to points near the end q, and vice versa. It follows by continuity that some point u on $\mathscr{L}$ is fixed by f, and hence the unique non-Euclidean line $\mathscr{M}$ through u and ending at r is mapped into itself by f. Also, because any Möbius transformation preserves angles, $\mathscr{M}$ must be perpendicular to $\mathscr{L}$ (Figure 8.14). Thus, *f has the same effect on p, q, r as reflection in the line $\mathscr{M}$*, so *f is* this reflection by "uniqueness of three-point maps."

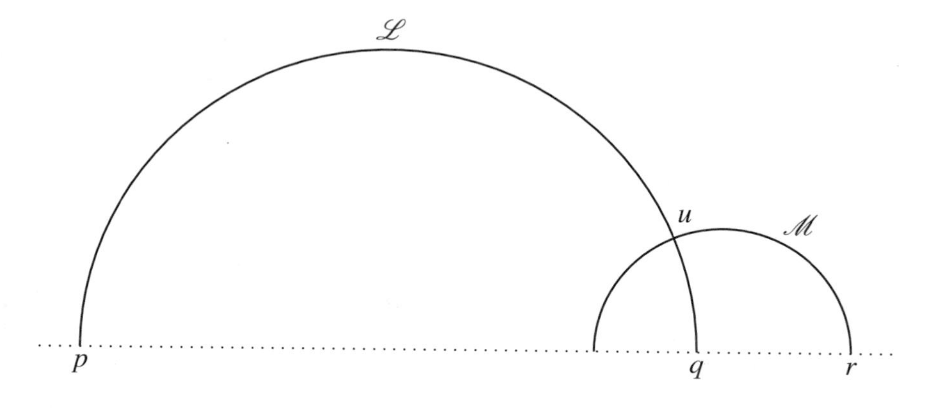

Figure 8.14: Lines involved in the involution f

Now consider the involution g, which is associated with a similar pair of lines $\mathscr{L}$ and $\mathscr{M}$. Only the names of their ends are different, but the reader is invited to draw them to keep track.

By the argument just used for f, the involution g is a reflection if it has a fixed point on $\mathbb{RP}^1$. In any case, g maps the line $\mathscr{L}$ with ends p and r into itself, exchanging the ends, so g has a fixed point u on $\mathscr{L}$ by the argument just used for f. Also, because g preserves angles, g maps the non-Euclidean line $\mathscr{M}$ through u and perpendicular to $\mathscr{L}$ into itself. Thus, if g has no fixed point on $\mathbb{RP}^1$, it necessarily exchanges the ends s and t of $\mathscr{M}$. But then *g has the same effect on the three points p, r, s as the product of reflections in $\mathscr{L}$ and $\mathscr{M}$*, so *g is* this product of reflections, by "uniqueness of three-point maps" again.

Thus, fg, which is an arbitrary Möbius transformation by the theorem above, is the product of at most three reflections. □

Exercises

The argument above appeals to "continuity" to show the existence of a fixed point on a non-Euclidean line whose ends are exchanged by an involution. This argument is valid, and it may be justified by the *intermediate value theorem*, well known from real analysis courses. However, some readers may prefer an actual computation of the fixed point. One way to do it is as follows.

Suppose $f(x) = \frac{ax+b}{cx+d}$ and that $f(p) = q$, $f(q) = p$.

8.8.1 Deduce that $a = -d$ and $b = cpq - a(p+q)$, so that f has the form

$$f(x) = \frac{a(x-p-q)+cpq}{cx-a} = \frac{k(x-p-q)+pq}{x-k} \quad \text{if } c \neq 0.$$

8.8.2 Solve the equation

$$x = \frac{k(x-p-q)+pq}{x-k},$$

and hence show that the fixed points of f are

$$u = k \pm \sqrt{(k-p)(k-q)}.$$

8.8.3 Assuming that $(k-p)(k-q) < 0$, so one fixed point is in the upper half plane, show that its distance from the center $(p+q)/2$ of the semicircle with ends p and q is $|(p-q)/2|$.

8.8.4 Deduce from Exercises 8.8.1–8.8.3 that f has a fixed point on the non-Euclidean line with ends p and q.

8.9 Discussion

The non-Euclidean parallel hypothesis

It has often been said that the germ of non-Euclidean geometry is in Euclid's own work, because Euclid recognized the exceptional character of the parallel axiom and used it only when it was unavoidable. Later geometers noted several plausible equivalents of the parallel axiom, such as

- the equidistant curve of a line is a line,
- the angle sum of a triangle is π,
- similar figures of different sizes exist,

but no outright proof of it from Euclid's other axioms was found. On the contrary, attempts to derive a contradiction from the existence of many parallels—what we will call the *non-Euclidean parallel hypothesis*—led to a rich and apparently coherent geometry. This is the geometry we have been exploring in the half plane, now called *hyperbolic geometry*.

Hyperbolic geometry diverges from Euclidean geometry in the opposite direction from spherical geometry—for example, the angle sum of a triangle is $< \pi$, not $> \pi$—but the divergence is less extreme. The "lines" of spherical geometry violate all three of Euclid's axioms about lines, whereas the "lines" of hyperbolic geometry violate only the parallel axiom.

The first theorems of hyperbolic geometry were derived by the Italian Jesuit Girolamo Saccheri in an attempt to prove the parallel axiom. In his 1733 book, *Euclides ab omni naevo vindicatus* (Euclid cleared of every flaw), Saccheri assumed the non-Euclidean parallel hypothesis, and sought a contradiction. What he found were *asymptotic lines*: lines that do not meet but approach each other arbitrarily closely. This discovery was curious, and more curious at infinity, where Saccheri claimed that the asymptotic lines would meet *and* have a common perpendicular. Finding this "repugnant to the nature of a straight line," he declared a victory for Euclid.

But the common perpendicular at infinity is *not* a contradiction, and indeed (as we now know) it clearly holds in the half plane. There are non-Euclidean lines that approach each other arbitrarily closely in non-Euclidean distance, such as the unit semicircle and the line $x = 1$, and they have a common perpendicular at infinity—the x-axis. Saccheri had unwittingly discovered not a bug, but a key feature of hyperbolic geometry.

The non-Euclidean geometry of the hyperbolic plane began to take shape in the early 19th century. A small circle of mathematicians around Carl Friedrich Gauss (1777–1855) explored the consequences of the non-Euclidean parallel hypothesis, although Gauss did not publish on the subject through fear of ridicule. Gauss was the greatest mathematician of his time, but he was unwilling to publish "unripe" work, and he evidently felt that non-Euclidean geometry lacked a solid foundation. He knew of no concrete interpretation, or *model*, of non-Euclidean geometry, and in fact, none was discovered in his lifetime. It is a great irony that some of his own discoveries—in the geometry of curved surfaces and the geometry of complex numbers—can provide such models.

The first to publish comprehensive accounts of non-Euclidean geometry were Janos Bolyai in Hungary and Nikolai Lobachevsky in Russia. Around 1830 they discovered this geometry independently and became its first "true believers." The richness and coherence of their results convinced them that they had discovered a new geometric world, as real as the world of mainstream geometry and not needing its support. In a sense, they were right, but in their enthusiasm, they failed to notice another new geometric theory that could have been a valuable ally. Gauss's *Disquisitiones generales circa superficies curvas* (General investigations on curved surfaces) was published in 1827, but neither Bolyai, Lobachevsky, or Gauss noticed that it gives models of non-Euclidean geometry, at least in small regions.

The fundamental concept of Gauss's surface theory is the *curvature*, a quantity that is positive (and constant) for a sphere, zero for the plane and cylinder, and negative for surfaces that are "saddle-shaped" in the neighborhood of each point. In the *Disquisitiones*, Gauss investigated the relationship between the curvature of a surface and the behavior of its *geodesics*, which are its curves of shortest length and hence its "lines." He found, for example, that a geodesic triangle has

- angle sum $> \pi$ on a surface of positive curvature,
- angle sum π on a surface of zero curvature,
- angle sum $< \pi$ on a surface of negative curvature.

Moreover, if the curvature is *constant* and nonzero, then, in any geodesic triangle, (angle sum $-\pi$) is proportional to area. These results must have reminded Gauss of things he already knew in non-Euclidean geometry, so it is surprising that he failed to capitalize on them.

Close encounters between the actual and the hypothetical

The near agreement between geometry on surfaces of constant negative curvature and non-Euclidean geometry was the first of several close encounters over the next few decades. But usually the actual and hypothetical geometries passed each other like ships in a thick fog.

For example, in the late 1830s, the German mathematician Ferdinand Minding worked out the formulas of negative-curvature trigonometry. He found that they are like those of spherical trigonometry, but with hyperbolic functions in place of circular functions. At about the same time (and in the same journal!), Lobachevsky showed that the same formulas hold for triangles in his non-Euclidean plane. This would have been a nice time to introduce the name "hyperbolic geometry" for the non-Euclidean geometry of constant negative curvature, but apparently neither Minding nor Lobachevsky realized that they might have been talking about the same thing. Perhaps they were aware of a difficulty with the known surfaces of negative curvature: They are *incomplete* in the sense that their "lines" cannot be extended indefinitely. Hence, they fail to satisfy Euclid's second axiom for lines.

The simplest surface of constant negative curvature is called the *pseudosphere* (somewhat misleadingly, because constant curvature is about all it has in common with the sphere). It is more accurately known as the *tractroid*, because it is the surface of revolution of the curve known as the *tractrix*. The defining property of the tractrix is that its tangent has constant length a between the curve and the x-axis (left half of Figure 8.15).

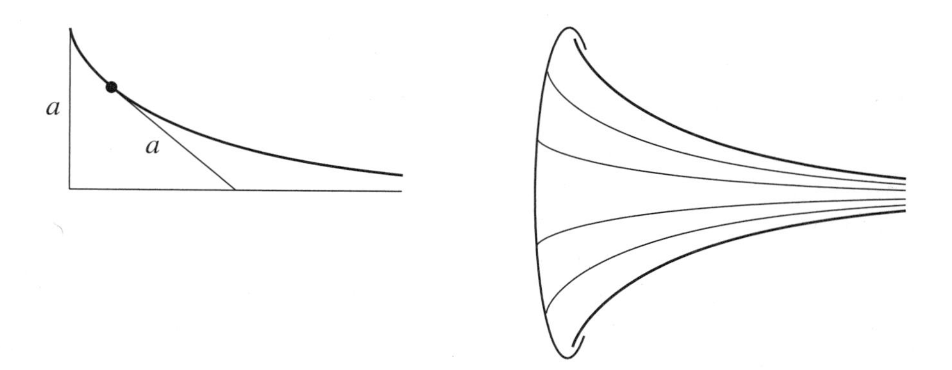

Figure 8.15: The tractrix and the tractroid

It is an unavoidable consequence of this definition that the tractrix has a singularity where the tangent becomes perpendicular to the x-axis. The tractroid likewise has an *edge* (like the rim of a trumpet), beyond which it cannot be smoothly continued. Hence, geodesics on the tractroid cannot be continued in both directions. In fact, the only geodesics on the tractroid that are infinite in even one direction are the rotated copies of the original tractrix. This problem is typical of what happens when one tries to construct a complete surface of constant negative curvature in ordinary space. The task was eventually shown to be impossible by Hilbert in 1901, but an obstacle to the construction of such surfaces was sensed much earlier.

In 1854, Gauss's student Bernhard Riemann showed a way round the obstacle by proposing an *abstract* or *intrinsic* definition of curved spaces—one that does not require a "flat" space to contain the "curved" one. This idea made it possible to define a complete surface, or indeed a complete n-dimensional space, of constant negative curvature. Riemann did exactly this, but once again non-Euclidean geometry sailed by unnoticed, as far as we know. (The elderly Gauss was very moved by Riemann's account of his discoveries. Whether he saw in them a vindication of non-Euclidean geometry, we will probably never know.)

Another close encounter occurred in 1859, when Arthur Cayley developed the concept of distance in projective geometry. He found that there is an invariant length for certain groups of projective transformations, such as those that map the circle into itself. In effect, he had discovered a model of the non-Euclidean plane, but he did not notice that his invariant length had the same properties as non-Euclidean length. Despite the efforts of Bolyai and Lobachevsky, non-Euclidean geometry remained an obscure subject until the 1860s.

Models of non-Euclidean geometry

Riemann died in 1859, and his ideas first bore fruit in Italy, where he had spent a lot of time in his final years. His most important successor was Eugenio Beltrami, who in 1868 finally brought non-Euclidean geometry and negative curvature together.

Beltrami's first discovery, in 1865, established the special role of constant curvature in geometry: *The surfaces of constant curvature are precisely those that can be mapped to the plane in such a way that geodesics go to straight lines.* The simplest example is the sphere, whose geodesics are great circles, the intersections of the sphere with the planes through its center. Great circles can be mapped to straight lines by projecting the sphere onto the plane from its center (Figure 8.16).

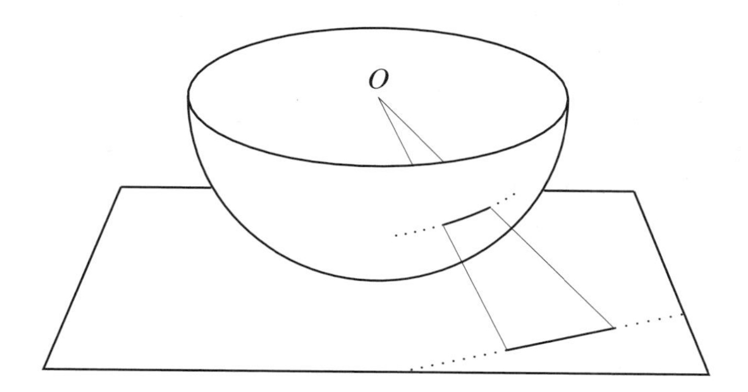

Figure 8.16: Central projection of the sphere

The geodesic-preserving map of the tractroid sends it to a wedge-shaped portion of the unit disk. The tractrix curves on the tractroid go to line segments ending at the sharp end of the wedge (Figure 8.17).

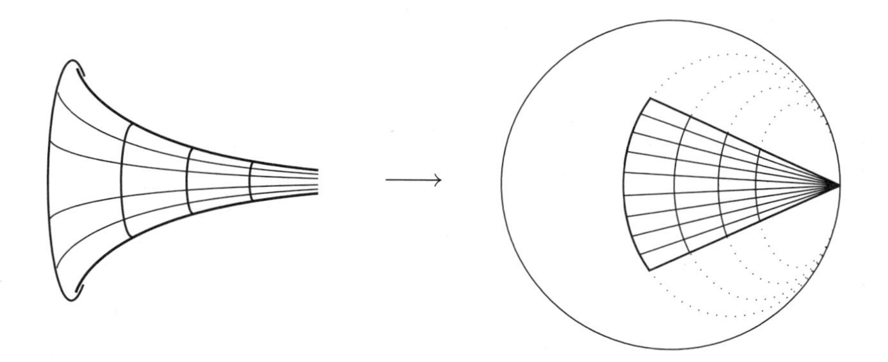

Figure 8.17: Geodesic-preserving map of the tractroid

Although this map preserves "lines," it certainly does not preserve length. Each tractrix curve has infinite length; yet it is mapped to a finite line segment in the disk. The appropriate length function for the disk assigns a "pseudodistance" to each pair of points, equal to the geodesic distance between the corresponding points on the tractroid. We do not need the formula here; the important thing is that *pseudodistance makes sense on the whole open disk*, that is, for all points inside the boundary circle. The curve of shortest pseudodistance between any two points in the open disk is the straight line segment between them, and the pseudodistance between any point and the boundary is infinite.

In 1868, Beltrami realized that this abstraction and extension of the tractroid is an *interpretation of the non-Euclidean plane*: a surface in which there is a unique "line" between any two points, "lines" are infinite, and the non-Euclidean parallel hypothesis is satisfied. Figure 8.18 shows why: Many "lines" through the point P do not meet the "line" $\mathscr{L}$.

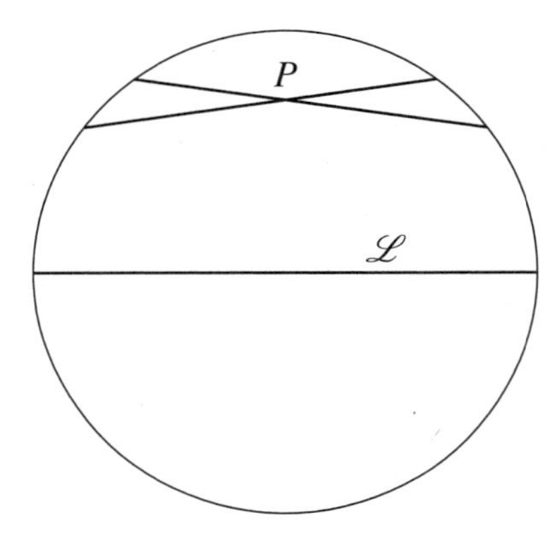

Figure 8.18: Why the non-Euclidean parallel hypothesis holds

Beltrami wrote two epic papers on models of non-Euclidean geometry in 1868, and English translations of them may be found in my book *Sources of Hyperbolic Geometry*. The first paper arrives at the non-Euclidean plane as an extension of the tractroid through the idea of "unwinding" infinitely thin sheets wrapped around it. (The dotted paths in the right half of Figure 8.17, all converging to the endpoint of the wedge, are the limit circles traced by circular sections of the tractroid as they unwind.) Beltrami was at pains to be as concrete as possible, because Riemann's ideas were not well understood or accepted in 1868. However, at the end of the paper, Beltrami foreshadows the more abstract and general approach he intends to take in his second paper:

> where the most general principles of non-Euclidean geometry are considered independently of their possible relations with ordinary geometric entities. In the present work we have been

> interested mainly in offering a concrete counterpart of abstract geometry; however, we do not wish to omit a declaration that the validity of the new order of concepts does not depend on the possibility of such a counterpart.

In the second paper, Beltrami vindicates this ringing endorsement of Riemann's ideas with whole families of models of non-Euclidean geometry in any number of dimensions. Among them is the half-plane model used in this chapter, and its generalization to three dimensions, the "half-space model." The half-space model has

- "points" that are the points $(x,y,z) \in \mathbb{R}^3$ with $z > 0$,
- "lines" that are the vertical Euclidean half lines in $\mathbb{R}^3$ and the vertical semicircles with centers on the plane $z = 0$,
- "planes" that are the vertical Euclidean half planes in $\mathbb{R}^3$ and hemispheres with centers on $z = 0$.

It turns out that non-Euclidean distance on a plane $z = a$ is a constant multiple of Euclidean distance. This surprising result gives probably the simplest proof of a result first discovered by Friedrich Wachter, a member of Gauss's circle, in 1816: *Three-dimensional non-Euclidean geometry contains a model of the Euclidean plane.*

Another model of the hyperbolic plane, discovered by Beltrami, is the *conformal disk model.* It is like the half plane in being angle-preserving, but unlike it in being finite. Its "points" are the interior points of the unit disk (the points z with $|z| < 1$, if we work in the plane of complex numbers), and its "lines" are circular arcs perpendicular to the unit circle. Figure 8.19, which is the original M. C. Escher picture *Circle Limit I*, can be viewed as a picture of the conformal disk model. The fish are arranged along "lines," and they are all of the same hyperbolic length. As mentioned in connection with Figure 8.10, the transformed *Circle Limit I*, the function

$$z \mapsto \frac{1 - zi}{z - i}$$

maps the conformal disk model onto the half-plane model.

It should be stressed that *all models of non-Euclidean geometry, in a given dimension, are isomorphic to the half-space model.* For example, models of the non-Euclidean plane satisfy Hilbert's axioms (Section 2.9)

Figure 8.19: The conformal disk model

with the parallel axiom replaced by the non-Euclidean parallel hypothesis. And Hilbert in his *Grundlagen* showed that the "lines" satisfying these axioms have "ends" that behave like the points of $\mathbb{RP}^1$. Thus, any non-Euclidean plane is essentially the same as the half plane discussed in this chapter, so we can call it *the* non-Euclidean plane or *the* hyperbolic plane.

Non-Euclidean reality

In Beltrami's original model, the open disk in which "lines" are line segments ending on the unit circle, isometries map Euclidean lines to Euclidean lines, and so they are projective maps. For this reason, the model is often called the *projective disk.* It can also be constructed by methods of projective geometry, and indeed this is essentially what Cayley did in 1859. The first to connect all the dots between projective and non-Euclidean geometry was Klein in 1871. An English translation of his paper may be found in *Sources of Hyperbolic Geometry.* Although Klein had only to fill a few technical gaps, it was he who first made the important conceptual point that *a model of non-Euclidean geometry ensures that the non-Euclidean parallel hypothesis is not contradictory. Hence, Euclid's parallel axiom does not follow from his other axioms.*

In 1872, Klein also made the great advance of linking geometries to groups of transformations. This link gives a deeper reason for the presence of non-Euclidean geometry in projective geometry: The real projective line and the non-Euclidean plane have isomorphic groups of transformations.

The group of the non-Euclidean plane was first described explicitly by the French mathematician Henri Poincaré in 1882, along with its interpretation as the group of Möbius transformations of the half plane.The relevant parts of his work may also be found in *Sources of Hyperbolic Geometry*. Poincaré became interested in non-Euclidean geometry when he noticed that some functions of a complex variable have *non-Euclidean periodicity*.

An ordinary periodic function, such as $\cos x$, has *Euclidean periodicity* in the sense that its values repeat when x undergoes the Euclidean translation $x \mapsto x + 2\pi$. A complex function can have non-Euclidean periodicity, and one example is the *modular function* $j(z)$. Its definition is too long to explain here, but its periodicity is simple: The values of $j(z)$ repeat under the Möbius transformations $z \mapsto z+1$ and $z \mapsto -1/z$. As we know, these are isometries of the half plane. If one applies them over and over, to the lines $x = 0$, $x = 1$, and the unit semicircle, they produce the non-Euclidean regular tessellation shown in Figure 8.20.

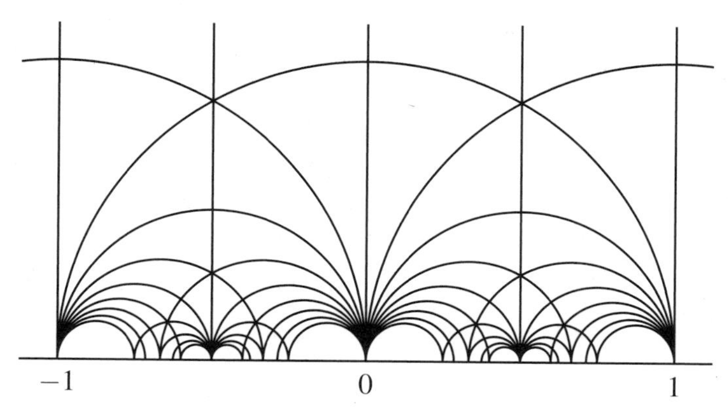

Figure 8.20: The modular tessellation

The modular function and its periodicity were already part of mathematical reality, having been known to Gauss and others since early in the 19th century. But Poincaré was the first to see its non-Euclidean symmetry. He used non-Euclidean geometry to study large classes of functions whose behavior had until then seemed intractable. Poincaré was also the first to view the half plane as an extension of the real projective line, as we have done in this chapter. In fact, he went much further, noticing that the half-space model of non-Euclidean space is a natural extension of the *complex* projective line $\mathbb{CP}^1 = \mathbb{C} \cup \{\infty\}$.

Just as the real projective line $\mathbb{R} \cup \{\infty\}$ comes with the linear fractional transformations

$$x \mapsto \frac{ax+b}{cx+d}, \qquad \text{where } a,b,c,d \in \mathbb{R} \text{ and } ad-bc \neq 0,$$

the complex projective line $\mathbb{C} \cup \{\infty\}$ comes with the linear fractional transformations

$$z \mapsto \frac{az+b}{cz+d}, \qquad \text{where } a,b,c,d \in \mathbb{C} \text{ and } ad-bc \neq 0.$$

And just as the linear fractional transformations of $\mathbb{R} \cup \{\infty\}$ extend to Möbius transformations of the half plane, the linear fractional transformations of $\mathbb{C} \cup \{\infty\}$ extend to Möbius transformations of the *half space*, for which there is likewise an invariant non-Euclidean distance, and the non-Euclidean "lines" and "planes" mentioned above.

It is a great advantage to have a concept of distance, even if the distance is non-Euclidean and one needs an extra dimension to acquire it. By passing to the third dimension, Poincaré could understand transformations of $\mathbb{C}$ whose behavior is almost incomprehensible when viewed in the plane. Understanding comes by viewing these transformations as compressed versions of isometries of non-Euclidean space, which behave quite simply (like isometries of the half plane). Thus, expanding from a projective line to a non-Euclidean space is not just an interesting theoretical possibility—it is sometimes the best way to understand the mysteries of projection.

References

Artmann, B.: *Euclid: The Creation of Mathematics*, Springer-Verlag, 1999.

Behnke, H. *et al.*: *Fundamentals of Mathematics, Volume II. Geometry*, MIT Press, 1974.

Brieskorn, E. and Knörrer, H.: *Plane Algebraic Curves*, Birkhäuser, 1986.

Cayley, A.: *Mathematical Papers*, Cambridge University Press, 1889–1898.

Coxeter, H. S. M.: *Introduction to Geometry*, Wiley, 1969.

Descartes, R.: *The Geometry*, Dover, 1954.

Ebbinghaus, H.-D. et al.: *Numbers*, Springer-Verlag, 1991.

Euclid: *The Thirteen Books of Euclid's Elements*, Edited by Sir Thomas Heath, Cambridge University Press, 1925. Reprinted by Dover, 1954.

Euclid: *Elements*, Edited by Dana Densmore, Green Lion Press, 2002.

Gauss, C. F.: *Disquisitiones generales circa superficies curvas*, Parallel Latin and English text in Dombrowski, P.: *150 Years after Gauss' "Disquisitiones generales circa superficies curvas,"* Société Mathématique de France, 1979.

Hartshorne, R.: *Geometry: Euclid and Beyond*, Springer-Verlag, 2000.

Hilbert, D.: *Foundations of Geometry*,
Open Court, 1971.

Hilbert, D. and Cohn-Vossen, S.: *Geometry and the Imagination*,
Chelsea, 1952.

Kaplansky, I.: *Linear Algebra and Geometry, a Second Course*,
Allyn and Bacon, 1969.

McKean, H. and Moll, V.: *Elliptic Curves*,
Cambridge University Press, 1997.

Pappus of Alexandria: *Book 7 of the* Collection, Parts 1 and 2,
Edited with translation and commentary by Alexander Jones,
Springer-Verlag, 1986.

Saccheri, G.: *Euclides ab omni naevo vindicatus*,
English translation, Open Court, 1920.

Snapper, E. and Troyer, R. J.: *Metric Affine Geometry*,
Academic Press, 1971.
Reprinted by Dover, 1989.

Stillwell, J.: *Geometry of Surfaces*,
Springer-Verlag, 1992.

Stillwell, J.: *Sources of Hyperbolic Geometry*,
American Mathematical Society, 1996.

Veblen, O. and Young, J. W.: *Projective Geometry*,
Ginn and Company, 1910.

Wright, L.: *Perspective in Perspective*,
Routledge and Kegan Paul, 1983.

Index

(continued from page ii)